RIVERSIDE COMMUNITY
1916

P9-DVU-921

PENGUIN BOOKS

# THE GREEN CONSUMER

JOHN ELKINGTON is one of Europe's leading authorities on the role of industry in sustainable development. He runs an independent consultancy whose clients have included British Petroleum, the Nature Conservancy Council, and the United Nations Environment Programme. Having written numerous books and reports, he is also the co-author, with Tom Burke, of *The Green Capitalists*.

JULIA HAILES has worked in advertising and TV production. In 1987, she helped to set up SustainAbility, "the green growth company," with John Elkington and Tom Burke. The company aims to promote environmentally sustainable economic growth.

JOEL MAKOWER is a Washington, D.C.-based writer and lecturer on consumer and environmental topics, and editor of *The Green Consumer Letter*, a monthly newsletter. He is a graduate in journalism from the University of California at Berkeley. A former consumer columnist, he is author or co-author of more than a dozen books.

JOHN ELKINGTON, JULIA HAILES,
AND JOEL MAKOWER

*A Tilden Press Book*

PENGUIN BOOKS

PENGUIN BOOKS
Published by the Penguin Group
Viking Penguin, a division of Penguin Books USA Inc.,
375 Hudson Street, New York, New York 10014, U.S.A.
Penguin Books Ltd, 27 Wrights Lane,
London W8 5TZ, England
Penguin Books Australia Ltd, Ringwood,
Victoria, Australia
Penguin Books Canada Ltd, 2801 John Street,
Markham, Ontario, Canada L3R 1B4
Penguin Books (N.Z.) Ltd, 182–190 Wairau Road,
Auckland 10, New Zealand

Penguin Books Ltd, Registered Offices:
Harmondsworth, Middlesex, England

First published in Great Britain by Victor Gollancz Ltd. 1988
This new, updated edition first published in
the United States of America in Penguin Books 1990

5 7 9 10 8 6 4



Research Director—Cathryn Poff
Associate Researcher—Deborah McLaren

Apple logo, © Victor Gollancz Ltd.

LIBRARY OF CONGRESS CATALOGING IN PUBLICATION DATA
Elkington, John.
The green consumer/John Elkington, Julia Hailes, and Joel Makower.
p. cm.
Includes bibliographical references.
ISBN 0 14 01.2708 9
1. Consumption (Economics)—United States. 2. Environmental protection—United States. I. Hailes, Julia. II. Makower, Joel, 1952– . III. Title.
HC110.C6E44 1990
363.7'057—dc20 89-48639

Printed in the United States of America

Set in Palatino

## About This Book

We have divided *The Green Consumer* into three parts. Part I offers the broad view of Green Consumerism, the state of the environment, and how your everyday purchases can affect the earth's resources. Part II features a comprehensive guide to what products to buy—and not to buy—including brand names and addresses. Part III is for those who want to learn more or become more involved in environmental issues, including names and addresses of dozens of environmental organizations.

There are two principal ways this book can be of use. One, of course, is through the listings of green products and companies. Understand, however, that a product's inclusion in this book does not constitute an endorsement of this product, nor does it ensure a product's quality and efficacy. The second and equally important use of this book is to obtain a better understanding that will help you to judge *all* products for their potential environmental impact. By learning to include environmental concerns along with price, quality, nutrition, and convenience, you will be very well equipped to meet the demanding task of being a Green Consumer in the 1990s and beyond.

When contacting any of the companies in this book, be aware that each has its own policies and practices. Some sell directly to consumers, accepting phone orders and credit card charges. Others accept orders by mail, accompanied with payment. (Note that prices listed in this book are suggested retail prices and do not include shipping or sales taxes.) Still other companies will refer you to a local retailer that carries their products. Some companies offer catalogs and brochures of their products, while others have no such information; although most of this literature is free, there may be a charge for some, or a requirement to send a stamped, self-addressed envelope. Moreover, all of these policies are subject to change. As this book went to press, for example, several companies that had not previously sold directly to consumers reported that they were considering doing so. Others were preparing—or thinking of preparing—brochures or catalogs for the first time.

We hope you find the information contained in these pages to be valuable. We welcome your ideas, suggestions, and comments, which we will incorporate in future editions of *The Green Consumer*. Please send them to The Green Consumer, c/o Tilden Press, Inc., 1526 Connecticut Ave. NW, Washington, DC 20036.

## AUTHORS' NOTE

The authors and the publishers of the American edition of this work have prepared this work and set forth the views contained in it in good faith.

The omission of any particular product, company, or any other organization from this work implies neither censure nor recommendation. Neither the authors nor the publishers of this edition of the work, however, warrant effectiveness or performance of any product set out in this work, and the reader and consumer must exercise his or her own judgment when determining what criteria to apply when judging whether or not he or she should purchase any particular product from any of those companies or other organizations which are mentioned in this work. Any reader requiring further information concerning any particular product should write directly to the supplier or manufacturer of that product.

The authors and the publishers are willing to consider any matter which is held to be inaccurate and upon satisfactory factual and documentary confirmation of the correct position will use all reasonable endeavors to amend the text for the next reprint or edition.

While every care is used and good faith is exercised in assembling and publishing the material in this book, no warranty as to the properties or safety of any product is imputed by the inclusion of any such product in this work, and consumers should address themselves to the manufacturers of any product in respect of which they may wish to be informed as to its qualities.

## Contents

## *Part III: How to Get Involved*

## Acknowledgments

A great many individuals and organizations helped in this effort, and they deserve recognition and thanks:

American Council for an Energy-Efficient Economy; American Paper Institute; Brian Bence, Greenpeace; Peter Beren, Sierra Club; Sue Borra, Food Marketing Institute; Dianne Brause, Lost Valley Center; Sara Clark, Environmental Defense Fund; Jason Clay, Cultural Survival; David H. Collins, Council on Economic Priorities; Liz Cook, Friends of the Earth; Pamela Dorman, Penguin Books; Ivan Eussach, Rainforest Alliance; George A. Everett, National Appropriate Technology Assistance Service; Jay Feldman, National Coalition Against the Misuse of Pesticides; *Garbage* magazine; Nora Goldstein, *BioCycle*; Sondra Goodman, Household Hazardous Waste Project; Claudie Grout, Environmental Hazards Management Institute; Suzanne Head, Rainforest Action Network; Alan Herschgawitz, Natural Resources Defense Council; Guy Hodge, Humane Society of the United States; John Javna, The Earth Works Group; Debbie Keller, New Jersey Environmental Federation; Jane Kochersperger, National Coalition Against the Misuse of Pesticides; Janet Kraybill, Penguin Books; Ruth Lampi, Fairfax County Division of Solid Waste; Annie Leonard, Greenpeace; Alice Tepper Marlin, Council on Economic Priorities; John A. May; Barbara McBride, Food Marketing Institute; National Wildlife Federation; Jerry Powell, *Resource Recycling Magazine*; Todd Putnam, *National Boycott Newsletter*; Thomas A. Robertson; Rocky Mountain Institute; Jonathan Schorsch, Council on Economic Priorities; Suzanne Schmidt, Natural Resources Defense Council; Drusilla Schmidt-Perkins, Energy Conservation Coalition; Ed Stana, Council on Plastic Packaging in the Environment; Linda Starke; Mary Stone, League of Women Voters; Kay Treakle, Greenpeace; Hawley Truax, Environmental Action; Mike Walsh; Rob Watson, Natural Resources Defense Council; Dana West, World Wildlife Fund; Simon Williams, The Michael Peters Group; World Resources Institute; Jeanne Wirka, Environmental Action; and Linda Worthington, Center for Science in the Public Interest.

## Foreword

*by Ben Cohen, Chairperson*
*Ben & Jerry's Homemade*

When Ben & Jerry's first started, I used to be a hot ticket on the Rotary Club speaker circuit. The assignment was to drone on a bit after lunch in order to help people digest their meal. Ben & Jerry's was a real small company at the time—we were a homemade ice cream parlor—and we used to do things for the community, like show free movies during the summer on the wall outside our gas station, and sponsor community celebrations, and give away free ice cream on our anniversary. Things like that.

I'd be talking to the Rotary Club about these things and at the end of the talk somebody would kind of lift up his head and say, "Well, you know, those things you're doing for the community—you're just doing them because it's good for business, right?" And I responded, "Well, I don't know, but our reason is that we genuinely believe that business has a responsibility to give back to the community, and we're doing it out of altruism."

That was my old answer. My new answer is, "Yeah, it *is* good for business. And if you're smart, you'll jump on the bandwagon."

I have been thinking a lot about the influencers of business. One of those influencers is capital—that is, the money that people invest in businesses by buying stocks. Many people are starting to invest their money only in companies whose values they agree with. Businesses need this investment capital in order to survive and prosper. So, as more and more investment capital has these kinds of strings attached to it, companies are starting to be influenced to go along with investors' values. Another influencer is the value of employees. Employees run and operate businesses. So, it makes sense to start educating business school students in the art and science of operating businesses in such a way that they make a profit and proactively support our communities—local, national, and global—at the same time. A third influencer is sales. Businesses need sales to exist, and if companies find that customers will support them more if they adopt a particular social stance, they are going to move in that direction.

That's all fine and good. But consumers need the tools with which to vote with their dollars and decide which businesses to support and which businesses have values that they agree with and which businesses don't. That's where being a Green Consumer comes in.

The beauty of it is that being a Green Consumer is a really easy thing to do. All you need is some education. You don't have to go out of your way in order to influence companies through your purchasing behavior. When you go into the supermarket, or any store for that matter, you usually have quite a few choices of brands to buy if you're looking for a particular product. In many cases we're talking about products between which there really isn't much difference. You simply choose which packaging you want or which color you want or which one has a slightly better price or a name that rings a bell in your head. Now, with this book, you can add another factor: Is this product and this company environmentally responsible?

Companies that want to survive into the twenty-first century know that they must consider values. At Ben & Jerry's, we operate as a values-led business, a term coined by Anita Roddick, founder of The Body Shop. We tend to pursue activities based on a particular set of values. So if our value is to create a world that spends less on the military, we will find a way to integrate that message into our day-to-day business activities and influence our government in that direction. If our value is to try to create less economic disparity in our society, we will support organizations that are working in that direction through our purchasing decisions, our investment decisions, and our internal salary structure. If our value is to try to help the environment, we will support that through internal company recycling programs, through externally trying to educate people about recycling, and by working with companies like Community Products, whose purpose is to prove that rain forests can be more profitable as living rain forests than as deforested land.

I think the real problem is that we need to redefine the bottom line for business success. The bottom line has to be in two parts. When we measure our success as a business at the end of the year, we must look at how much money is left over *and* how much of a contribution we made to the community—whether it's the local community or the nation or the world. We have to factor those two things together to determine how well we're doing.

As Green Consumers, you have the right and responsibility to vote—with your dollars—on how well businesses are doing these things: how successfully they are addressing the issues you believe are most important to your life, and to the life of our planet.

As the saying goes, vote early, and vote often.

# Part I

# T·H·E G·R·E·E·N CONSUMER

C·O·N·S·U·M·E·R

# I·N·T·R·O·D·U·C·T·I·O·N

You probably don't realize it, but every week you make dozens of decisions that directly affect the environment of the planet Earth. At work, at home, and at play, whether shopping for life's basic necessities, taking a vacation, or cleaning the house, the choices you make are a never-ending series of votes for or against the environment.

Buy a burger, fries, and a soda and you are probably worsening the already critical landfill crisis. Take your car in for repairs and you may be contributing to the gradual warming of the earth and increasing your chances of getting skin cancer. Do your laundry and you may be fouling America's lakes and rivers, perhaps your own drinking water. Discard your trash and you may be polluting the air, water, and soil, and helping to deplete the earth's natural resources.

But the products and services you buy need not be so destructive to the environment. By choosing carefully, you can have a positive impact on the environment without significantly compromising your way of life. That's what being a Green Consumer is all about.

It wasn't very long ago that being a Green Consumer was a contradiction in terms. To truly care for the environment, it was said, you had to drastically reduce your purchases of everything—food, clothing, appliances, and other "lifestyle" items—to a bare minimum. That approach simply doesn't work in our increasingly convenience- and consumption-oriented society. No one wants to go back to a less-comfortable, less-convenient way of life.

And yet most Americans do care about the state of the earth. Increasingly, according to surveys, people say that their concern for environmental issues is affecting the way they shop. More than half of the 1,000 adults in a 1989 survey conducted by the Michael Peters Group, an international marketing firm, said they chose not to buy a product in the last year because of concern that it or its packaging might harm the environment. Just over three-fourths said they would be willing to pay as much as 5 percent more for a product packaged with recyclable or biodegradable materials. Sixty-one percent of those in another survey said they were "much" or "somewhat" more inclined to patronize a store or restaurant that showed its concern over the environment by doing such things as reducing its use of plastic containers and utensils, and by recycling other waste materials.

Such concern notwithstanding, there's no question that the jump from environmental concern to environmental consumerism is easier said than done. As is true with so many other parts of our lives—dieting comes quickly to mind—one's good intentions don't always translate readily into effective action. In the case of being a Green Consumer, the problem stems in part from a lack of understanding about how your purchasing decisions can affect the environment, and about what qualities make your purchases "green."

## WHAT IS GREEN?

There are two basic ways in which a product can be considered green: it could have more environmentally sound contents, it could be wrapped in environmentally sound packaging—or both. Ideally, a green product is one that:

- ❑ is not dangerous to the health of people or animals
- ❑ does not cause damage to the environment during manufacture, use, or disposal
- ❑ does not consume a disproportionate amount of energy and other resources during manufacture, use, or disposal
- ❑ does not cause unnecessary waste, due either to excessive packaging or to a short useful life
- ❑ does not involve the unnecessary use of or cruelty to animals
- ❑ does not use materials derived from threatened species or environments

In addition, a green product ideally should not trade price, quality, nutrition, or convenience for environmental quality.

**The Many Shades of Green.** Meeting all these requirements is no small task, although a growing number of companies large and small are finding ways to meet the challenges. There are hundreds of green products introduced each year, with thousands more coming in the foreseeable future.

Unfortunately, few of these products are perfectly green. Most incorporate some improvements in packaging or contents, but do not necessarily meet all of the criteria listed above. One big problem is that there is considerable disagreement even among dedicated environmentalists about whether some purportedly green products truly are less harmful to the environment. For example, is biodegradable plastic a suitable alternative to nonbiodegradable plastic? Some people strongly object to biodegradable plastic because it does not completely break down into benign materials when disposed of in landfills. Others applaud the use of biodegradable plastic, pointing to the fact that while not a perfect solution, it is at least one step better than the nonbiodegradable variety. (More on the debate over biodegradable plastic in "The Problem of Packaging.") Is it better to do something imperfect now or wait for perfect solutions to come later on? The answer is for you to decide.

Equally frustrating is that there are no nationally accepted standards or coding systems for determining what products are environmentally sound. There are no agreed-upon definitions of when something may rightfully be labeled as "biodegradable," "degradable," "recyclable," or "made from recycled materials," among the more common terms now being used on product labels. And there are even less-specific labeling terms, such as "not harmful to the environment" or "environment-friendly."

The result is a mixed bag of green products. There are some environmentally harmful products wrapped in green packaging. Some green products don't clearly state their greenness, while other products claiming to be green are not. To make matters worse, several of the corporations producing green products are among the world's biggest polluters. In short, it's a confusing world, with many shades of green.

### *How the American Way of Life Is Destroying the Earth*

Per-person daily household trash produced in Calcutta, India:
1.12 pounds; in New York City: 3.96 pounds

Barrels of oil wasted annually because the federal government has not raised efficiency standards for cars by 1 mile per gallon:
420,000

Amount of oil the U.S. would have to import to meet present demand if the average fuel efficiency of all cars on U.S. roads averaged 42 MPG:
none

Pounds of agricultural pesticides applied each year in California:
80 million pounds

Portion of the 35,000 pesticides introduced since 1945 tested for potential health effects:
10 percent

Plastic beverage bottles Americans go through every hour:
2.5 million

Styrofoam cups thrown away each year in the U.S.:
25 billion

**The Power of Green Consumerism.** In the chapters that follow, we will attempt to lead you through this sometimes murky world of Green Consumerism. We have tried to present the different sides to some controversial issues, but it will be up to you to make the final decisions.

While those decisions won't always be easy to make, we urge that it is better to do something than nothing. While perfect solutions are still lacking, there are many companies making some

Americans living in areas with levels of air pollutants the federal government considers harmful:
110 million

Trees wasted each week by Sunday newspapers not being recycled:
500,000

Homes that could be heated by the wood and paper thrown away each year: 5 million homes for 200 years

Scrap tires generated by American drivers in 1988:
246.9 million

Plastic containers dumped overboard daily by commercial fishing fleets:
640,000

Northern fur seals drowned each year by lost plastic fishing net:
500,000

Estimated number of sea birds, marine mammals, and fur seals killed each year as a result of eating or being strangled by plastics:
1 million, 100,000, and 50,000, respectively

Gallons of water that can be contaminated by a single quart of motor oil:
up to 2 million

Grazing area required to produce a single all-beef hamburger:
55 square feet

attempt to improve the environmental quality of their products. Your support of these progressive companies and products will be heard loud and clear in the executive suites of the nation's largest companies, and will encourage other companies to follow these leaders.

You may be surprised at how easy it is to make your voice heard in the marketplace. The marketplace is not a democracy; you don't need a majority opinion to make change. Indeed, it takes only a fairly small portion of shoppers—as few as one person in ten—

## *How to Be a Green Consumer*

When it comes to evaluating and choosing green products—those that cause minimal impact on the environment—here are ten basic guidelines you should consider when evaluating a product:

- ❑ Look for products packaged in recyclable materials such as cardboard or glass, not in plastic containers.
- ❑ Don't buy products that are excessively packaged or wrapped.
- ❑ Look for products in reusable containers, or for which concentrated refills may be purchased.
- ❑ Look for products made from recycled paper, aluminum, and other reusable materials.
- ❑ Choose products that don't contain bleaches or dyes.
- ❑ Avoid buying anything packaged in Styrofoam or similar flexible foam materials.
- ❑ Don't confuse "green" with "healthy"; not everything packaged in recyclable packaging is necessarily good for you or the environment.
- ❑ Remember that there are few perfectly green products.
- ❑ Keep in mind that doing something is better than doing nothing.
- ❑ Carry your products home in paper, not plastic, bags. Better yet, bring your own cloth bag.

changing buying habits for companies to stand up and take notice. For example, if 15 percent of American consumers decided to buy only those napkins, paper towels, toilet paper, and other goods made of recycled paper, you can be sure that the nation's leading paper goods companies would seriously consider making recycled paper products widely available. Such dramatic changes in corporate behavior begin with changes in consumer behavior, one purchase at a time.

There are other things you can do to make your voice heard even louder. Consider the supermarket, one of the front lines in the battle for a better environment. A simple request to the manager of your local supermarket—or better yet, a call or letter to the regional manager or corporate headquarters—stating your preference for environmentally safe products will encourage them to stock such items. You may also suggest that the store offer shelf labeling, with signs (green, of course) highlighting green products. Stores in other countries have already heeded their customers' wishes. In Canada and several European countries, for example, supermarkets distribute their own green product lines.

Persuade one major company to change its tack and others are likely to follow. McDonald's decision to abandon the use of CFCs in fast-food cartons in the United States was one of the environmental milestones of 1987. The threat that consumers might boycott Big Macs was surely a factor influencing the company's decision. Competitors promptly started talking to their carton suppliers, explaining: "We don't want to be left behind."

There are other victories. For example, at its customers' urging, Wal-Mart, the national discount store chain, launched a national advertising campaign featuring full-page newspaper ads challenging its suppliers to provide products packaged in environmentally sound ways. Procter & Gamble, the huge packaged-goods company, has responded to customer outcry about the trash produced by Pampers, its disposable diaper, by launching a recycling experiment in Washington State. The company also is test-marketing household cleaners in concentrated form, packaged in plastic pouches, permitting customers to refill empty bottles by adding water instead of purchasing a new container. In Canada, Cascade-Dominion, one of the country's leading producers of foam egg cartons made from a material that harms the ozone layer, cited consumer pressure as the key motivator when it announced plans to close its foam egg carton plant and switch all production to pulp egg cartons.

The bandwagon effect will be inevitable: as markets and product manufacturers get the message that Americans want their products green, the trickle of green products will turn into a flood.

Please understand that buying green products alone won't solve the huge environmental problems facing our nation and our world. Many of the problems are beyond our individual control. Acid rain, for example, comes largely from antiquated power plants emitting sulfur dioxide and nitrogen oxides; solutions to this and other environmental problems must come principally from businesses and governments working together.

But there are many environmental problems that you *can* do something about. Being a Green Consumer is a good and effective start—your own personal message to Mother Earth that you are grateful for her many gifts.

## Seven Environmental Problems You Can Do Something About

You've probably been hearing for years about "the environment." But how much do you really understand the earth's environmental problems? A basic working knowledge of the biggest environmental concerns is the first step to being a Green Consumer. By understanding the problems, you will best be able to understand how your consumer purchases may be contributing to them.

While a technically thorough explanation of environmental causes and effects could fill an entire book (and there are many good books on these subjects; see "The Green Bookshelf"), below, in no particular order of importance, is a summary of the biggest environmental problems of the 1990s.

One of the things you will notice is that many of the problems are related to one another. Some of the same pollutants that contribute to global warming and the greenhouse effect also contribute to acid rain. Some of the same products that are contributing to the depletion of the ozone layer also contribute to the landfill crisis. Everything, it seems, is linked to something else. The beauty of nature is a delicate balancing act, and human pollution can throw just about everything into disarray.

Also note that while we have for years examined most of these problems within our own country's borders, the problems have become global. We have come to understand that our actions in one part of the world can affect the quality of life in another. Similarly, many solutions, to truly be effective, must come from an international effort.

Many of the solutions to the problems described below are covered more thoroughly in Part II of this book, which offers specific suggestions, along with product names and company addresses, that can help you do your part to minimize many of these problems. Also note that omitted from the "What You Can Do" sections below are activities of the citizen-action variety. In Part III, "How to Get Involved," we'll offer a variety of suggestions and resources to help you to take action on many of these environmental issues, including information on how to report polluters, how to press for legislation, organizations you can join, and strategies for making your voice heard.

## 1. Acid Rain

We have usually considered rain to be a cleansing experience. A good rainfall is thought to "clear the air" and give everything a nice fresh smell. But in certain parts of the United States as well as in other parts of the world, air pollutants mix with rainfall to pollute rivers and lakes and kill trees. Acid rain has been known to peel the paint off cars and is suspected of damaging the Canadian maple tree population so severely that the maple syrup industry there may be defunct by the turn of the century. Worst of all, acid rain is also known to represent a major health hazard to people.

The two biggest pollutants that contribute to acid rain are sulfur dioxide, which comes primarily from electric utilities' burning of high-sulfur coal, and nitrogen oxides, which also result from utility combustion, as well as the engines of cars and trucks. Once in the air, sulfur dioxide and nitrogen oxides combine with other airborne chemicals and water to form sulfuric acid and nitric acid, and undergo further chemical reactions to become sulfates and nitrates. When these chemicals mix with rain, sleet, snow, or hail, they fall to earth, where they can wreak havoc on just about everything. Even during dry weather, these acidic particles or gases can form on the ground, where they mix with soil or are taken up directly by plants.

One of the biggest problems of acid rain is the acidification it causes in lakes and streams, killing fish and other aquatic life. In the Adirondack Mountains of New York, numerous clear lakes believed to have supported fish life at one time have been found to be devoid of fish. The acidity levels of these lakes is considerably higher than those of nearby lakes that still contain fish. In addition, acid rain affects drinking water by causing several toxic materials to leach into drinking water: aluminum (which has been connected with Alzheimer's and Parkinson's diseases and central nervous system disorders), asbestos (a known cause of lung cancer and other respiratory diseases), cadmium (associated with kidney damage), and lead (associated with brain damage in children and increased risk of hypertension and heart disease in adult males). These toxins are picked up from corroded water pipes, the solder that holds the pipes together, and from whatever soil the water passes through before reaching reservoirs or seeping into underground aquifers—including city dumps and toxic landfills.

But most important, acid rain is being increasingly seen as a major threat to human health. For one thing, sulfur dioxide and nitrogen oxides, the chemicals that cause acid rain, themselves can harm the respiratory system. Most vulnerable are the very young and the elderly, particularly those already affected by asthma and bronchitis. Also at risk are pregnant women and people with heart disease. According to Dr. Philip J. Landrigan, a professor of community medicine and pediatrics at the Mount Sinai School of Medicine, acid rain may be the third largest cause of lung disease, after smoking and passive smoking. A congressional study blamed acid rain for contributing to some 50,000 premature deaths a year in the United States and Canada.

**What You Can Do.** There are limits to what individuals can do to control acid rain. However, there is at least one major solution:

❑ **Conserve energy.** Much of the problem results from industrial emissions, especially electricity-generating utilities that use high-sulfur coal. Reducing the need to build additional power plants is one effective means of easing the problem. At home, energy-wasting appliances are a major and needless drain on our resources. Switching to energy-efficient appliances not only saves money over the long term but also drastically reduces energy utility needs. According to the American Council for an Energy-Efficient Economy, if every household in the U.S. had the most energy-efficient refrigerators currently available, the electricity savings would eliminate the need for about ten large power plants.

## 2. Global Warming and the Greenhouse Effect

The world is warming, say the experts, and it is directly the result of the gases and pollutants we spew into the atmosphere. No one knows exactly how much the earth is heating up, how fast this is happening, or what the effects will be. But this much is known: Climatic zones are shifting, glaciers are melting, and the levels of our oceans are rising. By most estimates, as warming continues, forests will die, coastal areas will flood, the world's agricultural areas will wither, and there will be great economic upheaval.

Why are these disasters happening? Since the beginning of time, the sun has been beating down on the earth, providing warmth and light. Once it reaches the earth, that heat bounces back into space. If it didn't—if some or all of the heat were trapped inside the earth's atmosphere—the planet would gradually become a warmer place.

That's essentially what is happening. Since the dawn of the industrial revolution, we have been pumping billions of tons of carbon dioxide into the atmosphere, as well as smaller amounts of gases such as chlorofluorocarbons (CFCs), methane, and nitrous oxide. Those gases trap the sun's heat in the earth's atmosphere, just as a greenhouse traps warm air inside, keeping things warm even on a wintry day. So, the phenomenon has come to be called the greenhouse effect.

The greenhouse effect has serious implications for future generations. With the use of computer models, scientists have predicted that average global temperatures will increase between 3 and 10 degrees Fahrenheit by the middle of the next century. To put that into perspective, the average global temperature has not varied more than 4 degrees Fahrenheit in the 15,000 years since the last glacial period. According to the Union of Concerned Scientists, "If the projections of the greenhouse experts come true, weather can be expected to change in ways *beyond the range of our experience*, bringing extreme abnormalities of heat, cold, drought, and flood."

Reports of the effect of global warming—and the resulting rise in ocean levels due to the melting polar ice caps—are almost too sensational to seem real, but scientists say the scenarios are very real indeed. If water levels were to rise just 15 feet, much of Florida would end up under water; in Washington, D.C., water would inundate National Airport and the Lincoln Memorial and would nearly reach the Capitol steps.

Some of the greenhouse gases occur naturally. Without these gases, there would be no trapping of heat and the earth would be considerably colder and as lifeless as the moon. But the bulk of the greenhouse gases are the result of human activity. For example:

❑ **Carbon dioxide,** responsible for about 50 percent of the greenhouse effect, is created principally by the burning of fossil fuels—wood, coal, oil, and natural gas. Each year, human activity sends some 5.5 billion tons of carbon dioxide into the atmosphere. In the

United States, we generate about 6 tons of carbon dioxide per person annually; since 1986, carbon dioxide emissions have grown faster in the United States than in the rest of the world. Another source of carbon dioxide is the burning of trees, particularly the massive burning in tropical rain forests; see #5 below.

❑ **Chlorofluorocarbons (CFCs),** which account for roughly 15 to 20 percent of the problem, are industrial chemicals manufactured for use as coolants (for refrigerators, air conditioners), cleaning solvents (in computer manufacturing), plastic (as Styrofoam, among many other materials), and foam insulation. CFCs are the principal material responsible for the destruction of the atmospheric ozone layer (see #3 below).

❑ **Methane,** which contributes about 20 percent of the greenhouse effect, is produced by landfills when organic waste breaks down, as well as in production of livestock (especially plant-eating animals) in farming due to the effects of nitrogen-based fertilizers, and in rice paddies.

❑ **Nitrous oxide,** accounting for about 10 percent of the greenhouse problem, is formed when certain fertilizers break down in the soil, and by the burning of coal, oil, and other fossil fuels. This substance, which in certain concentrations is known as "laughing gas," is also known to deplete the atmospheric ozone layer.

In addition, other substances, including ground-based ozone smog, water vapor, and airborne particles, contribute to the greenhouse effect.

**What You Can Do.** There are several things individuals can do to help stem the tide of the greenhouse effect. Among them:

❑ **Conserve energy.** Because fossil fuels are the primary cause of the greenhouse effect, cutting energy is the primary solution. Insulate your home, buy energy-efficient appliances, buy fuel-efficient automobiles (and keep them tuned), and improve lighting efficiency; lighting and heating alone account for half of all our energy use.

❑ **Eliminate CFC pollution.** Reducing chlorofluorocarbons helps with another problem as well: the depletion of the ozone layer (see #3 below).

❑ **Reduce garbage.** Recycling and precycling (see "The Problem of

Packaging") not only reduces methane emissions but also eliminates the need for polluting incinerators; it also saves energy.

❑ **Plant a tree.** Every tree takes a certain amount of carbon dioxide out of the air and supplies us with healthy oxygen. Trees also filter out toxins that seep into soil, and prevent them from polluting ground water. If properly planted, trees cool urban neighborhoods, reducing the need for air conditioning. Planting bushes, ground cover, and even window boxes also can help. (See "How Trees Save the Earth" for more on planting trees.)

❑ **Buy organic food.** Organic farmers do not use nitrogen-based fertilizers, which have been found to contribute to methane levels in the atmosphere.

## 3. Ozone Depletion

You've probably become aware in recent years that a sunburn is not just painful—it can lead to skin cancer several years down the road. What would it be like if your risk of skin cancer was increased *every time you went outside*?

That is one of the likely problems resulting from the depletion of the ozone layer of the stratosphere, a thin layer of gas located between 12 and 30 miles above the earth's surface that shields the earth from most of the sun's radiation. Through a seemingly endless list of everyday products and processes, each of us is unwittingly advancing toward the day when the sun's natural shield will no longer be effective.

The stratospheric ozone *depletion* problem is commonly confused with the ground-level ozone *pollution* problem. In the latter problem, ground-level ozone is created when gases from automobile exhaust and other sources mix with sunlight to form photochemical smog. This ozone is the one that causes eye irritation and respiratory problems and that has made life unpleasant for millions of people in and around Los Angeles, among several other cities. Clearly, ground-level ozone is undesirable; we want ozone in the upper atmosphere, not in the air we breathe.

What causes stratospheric ozone depletion? The major known culprits are CFCs—particularly CFC-11 and CFC-12—which have been used in aerosols, in the manufacturing of certain types of plastics, foam insulation, and as coolants in refrigerators and air conditioners. These chemicals, whose emissions have grown rapidly since their introduction several decades ago, break apart when interacting with sunlight in the upper atmosphere, releasing chlorine atoms that destroy ozone molecules. Every chlorine atom ultimately destroys several thousand ozone molecules. The rate of ozone-destroying chlorine atoms in the stratosphere is now four to five times higher than normal and is increasing at a rate of about 5 percent a year. Even if we were to stop emitting CFCs today, the continuing reactions of existing chemicals would ensure the continued destruction of the ozone layer for at least a century.

One of the biggest sources of CFCs is polystyrene, also known by the brand name Styrofoam, a product of the Dow Chemical Company . In the production of Styrofoam, about 90 percent of the CFCs blown into polystyrene to make it foam are released into the atmosphere. CFCs continue to be released during the storage, use, and disposal of the finished foam products. Aside from their effect on the ozone layer, polystyrene products present a major disposal problem—they are either incinerated, dumped in landfills, or strewn about as litter. Incineration produces toxic air pollution and hazardous ash. Lack of landfill space is already a crisis (see #6 below) and chemicals from plastics already in landfill often seep into nearby groundwater. Polystyrene (and other plastic) litter not only is unsightly but when ingested by wildlife mistaking it for food causes thousands of deaths through ulceration, malnutrition, and suffocation.

According to Jay D. Hair, president of the National Wildlife Federation, a single polystyrene cup contains one billion billion (1,000,000,000,000,000,000) molecules of chlorofluorocarbon. "Each of those molecules will last about 150 years before it breaks down," says Hair. "It will take each of those molecules about 15 years to rise to the 25-mile level of the stratosphere where the protective ozone layer exists. When broken apart by ultraviolet radiation, each of those molecules will destroy 100,000 molecules of ozone. Think of it like a Pac Man video game, where you see those little Pac Men going around gobbling things up. That's what chlorofluorocarbons do to the protective stratospheric ozone layer."

Another source of CFCs are air conditioners, particularly those in cars, which use CFCs in their coolants. Some 90 million cars and light trucks in the United States have air conditioning. When an air conditioner is serviced or when the car or air-conditioning unit is disposed of, the CFC-laden coolant is released into the atmosphere. The Environmental Protection Agency has estimated that CFCs in cars alone accounts for as much as 25 percent of CFC use in the United States; lost coolant from cars account for an estimated 16 percent of ozone destruction. A growing number of service stations and repair shops are equipped with devices that capture the coolant without releasing it into the atmosphere; the captured coolant can then be recycled or properly disposed of. (See "Cars and CFCs" for more on this.)

Less known but far more destructive to the ozone than CFCs are halons, which are used extensively in fire extinguishers. While not as common as CFCs, halons may have 10 times the ozone depletion potential of the most damaging CFCs. Halon-1211 is used in portable fire extinguishers for industrial locations and in hand-held units for homes and offices. Halon-1301 is mainly used as a fire extinguishing agent, in portable extinguishers in aircraft, and in some sprinkler systems. Most halon emissions do not occur during fire extinguishing, but in testing of the equipment.

**What You Can Do.** First and foremost is to stop using foam products that contain CFCs. This is easier said than done. Styrofoam products alone number in the thousands—everything from coffee cups to egg cartons to supermarket meat trays. CFCs are either made with or are contained in automobile dashboards, bicycle seats, carpet pads, dusters for camera lens, electronic switches, furniture cushions, gum remover, home heating thermostats, ice machines, jewelry, and many other products. (See page 55 for a more complete listing.)

Fortunately, there is a growing list of substitutes for CFCs and halons. Despite the fact that the aerosol spray industry predicted the demise of the aerosol in 1978 when companies were required to eliminate CFCs from their products, few aerosol spray products actually disappeared; their manufacturers were able to find other propellants. Similarly, most indications suggest that there are other suitable substitutes for most CFC applications.

Among the other things you can do to avoid CFCs:

☐ **Avoid foam packaging.** Avoid buying fast food packaged in CFC-foam containers for drinks, sandwiches, or other menu items. Similarly, limit your purchases of grocery products packaged in Styrofoam.

❑ **Recycle coolants.** Patronize automobile and appliance service outlets that use devices that capture CFC coolant for recycling or proper disposal. Request that your own service station acquire such a device.

❑ **Use CFC-free insulation.** When adding insulation to your home, don't use rigid foam unless you can verify that it is CFC-free. Good alternatives are fiberglass and cellulose.

❑ **Use CFC-free fire extinguishers.** If you buy a fire extinguisher for your home, purchase a dry-chemical or sodium bicarbonate model; most others contain halons.

## 4. Air Pollution

Although Americans have expressed concern about air pollution for more than two decades, things aren't necessarily getting better. In 1989, the federal government issued the first ever national survey of toxic chemicals released into the air by industry, which showed that these chemicals are being emitted at rates that threaten public health. And the problem isn't limited to big cities like New York and Los Angeles: Not long ago, the Environmental Protection Agency cited a nearby coal-fired power plant as the culprit behind the bright white haze obscuring the view of the Grand Canyon. In Florida's Everglades—10,000 square miles of delicate marshland that is home to some of the country's rarest birds and other animals—pollution has made the fish unsafe for eating. The Wilderness Society has predicted that of all the national parks, the Everglades is the closest to extinction. Other parks, including Utah's Canyonlands National Park and Maine's Moosehorn wilderness area, are also plagued by pollution.

Among the biggest threats to human health from the air is ground-level ozone, formed when sunlight interacts with nitrogen oxides and hydrocarbons. Nitrogen oxides come primarily from power plants and automobiles; hydrocarbons are emitted by cars, oil-based paints, gasoline (particularly when gas vapors escape while you are filling your tank), dry cleaners, chemical factories, and other sources.

Ozone is used commercially as an industrial bleach. At full strength, it can dissolve concrete and rubber. Even at microscopic levels, ozone irritates breathing passages, causing shortness of breath, coughing, and throat irritation even in healthy people. For children, the elderly, and those with asthma and bronchitis, ozone can be a menace. Ozone is also toxic to crops and trees and may be related to the decline in forests in the eastern United States.

While many of our activities contribute to the problem—with automobile exhaust leading the way—one big villain is American industry, which dumps billions of pounds of toxins into the air each year. The EPA has identified 320 toxic chemicals in the air, only 7 of which are regulated by federal law; 60 of the chemicals have been identified as causing cancer. Most of these are used in various manufacturing processes, including carbon tetrachloride, butadiene, acrylonitrile, and cadmium.

But ordinary citizens, too, increasingly are being conscripted to join the battle for clean air. In Los Angeles, the city with the nation's most polluted air, officials have taken drastic action. The sweeping twenty-year plan adopted in 1989 will not only dramatically curtail automobile use but will also convert most vehicles to the use of nonpolluting fuels by the year 2009. That's just the beginning. The 5,500-page plan also calls for the banning of all aerosol hair sprays and deodorants (not just those containing chlorofluorocarbons), requires companies to install the best anti-smog equipment available regardless of cost, forces employers to encourage car pooling, and may eventually ban outdoor barbecues and gasoline-powered lawn mowers.

We may eventually see similar measures in other polluted and car-choked cities and areas, such as Atlanta, Baltimore, Chicago, suburban Connecticut, Dallas, Denver, Houston, Milwaukee, New York, Philadelphia, Phoenix, San Diego, San Francisco, and Washington, D.C. Automobiles in these and other places account for 40 percent of urban smog. (Another 40 percent comes from dry cleaners, which use solvents that emit fumes, and from bakeries, which release yeast by-products that sunlight changes to ozone.) Cars also emit nitrous oxides, a source of acid rain. While cars have become far less polluting than twenty years ago, there are 55 percent more cars on the road than in 1970, traveling more net miles.

Even the seemingly innocent act of filling your car with gasoline causes serious pollution problems. Among the vapors emitted during filling is benzene, a chemical that causes leukemia and blood disorders. About 80 percent of airborne benzene is believed to be released at gasoline pumps as vehicles are being filled or later when they are running. Moreover, when you "top off" your tank—that is, fill the tank as far to the top as possible—you are causing an additional problem: When the gas is heated, it expands and some of it spills out, dumping even more fumes into the atmosphere.

**What You Can Do.** There several ways to reduce your contribution to air pollution, most of which have to do with automobile maintenance and use.

❑ **Don't top off your tank.** When you fill your car with gasoline, fill it only until the pump's automatic shut-off mechanism engages.
❑ **Use vapor controls when filling your car.** If possible, fill your car at pumps that use devices that capture gas vapors (these are required at gas stations in California and the District of Columbia).
❑ **Keep your car in shape.** According to the Environmental Protection Agency, cars get about 40 percent better gas mileage when tuned compared to when untuned. Moreover, a tuned car emits 42 percent fewer hydrocarbons and 47 percent less carbon monoxide.
❑ **Drive as little as possible.** Obviously, the less you drive, the less you pollute. Whenever possible, use car pooling and public transportation.
❑ **Conserve energy.** Use of energy-efficient appliances and lighting will reduce the running of polluting power plants.
❑ **Use natural gas.** If your furnace is capable of operating with natural gas, you will be helping to cut back on three major air pollutants: carbon dioxide, carbon monoxide, and sulfur dioxide.

## 5. Rain Forests and Biodiversity

The world is losing trees—indeed, entire forests—at an astonishing rate. And as the trees disappear, so do thousands of species of other living things. And that may be the least of our problems. The disappearance of trees has been implicated in worsening the greenhouse effect and in air pollution in general.

The most devastating deforestation has taken place in the world's tropical rain forests. Occupying only 6 percent of the earth's surface, these forests are found in warm areas where

## *The Miraculous Deeds of Rosy Periwinkle*

Jay D. Hair, president of the National Wildlife Federation, tells a very personal tale that brings home the value of rainforests:

"About four years ago . . . my older daughter Whitney was very ill with cancer. She literally came within a few days of death. She is here today and she is a beautiful, 14-year-old, healthy, completely cured young lady. Why? The drug that saved her life was derived from a plant called the rosy periwinkle. The rosy periwinkle was a plant native to the island country of Madagascar. The irony of this personal story is that ninety percent of the forested area of Madagascar has been destroyed. One hundred percent of all native habitat of the rosy periwinkle is gone forever. And just at a time when we're learning about the marvels of biotechnology. We are losing entire genetic stocks of wild living resources at a time when we're learning about the potential medical marvels of some of these plants, like the one used to cure my daughter. We are destroying them and their potential values forever. This is a tragedy with incredible consequences to the future of global societies."

rainfall is 200 centimeters (about 80 inches) or more each year. This warm, moist environment allows tall, broad-leaved trees to flourish. The trees allow little sunlight onto the forest floor. It is believed that the majority of the earth's species lives in these unique environments, in particular insects and flowering plants.

Tropical forests are being permanently destroyed at an estimated rate of 27 million acres each year, mostly through wholesale burning designed to clear land for agricultural use—in particular, the raising of beef cattle for export chiefly to North America for fast-food hamburgers. (See "Fast Food and the Environment.") An additional 13 million acres of tropical forest are logged, usually in a fashion guaranteed not to ensure the forests' future. To put these 40 million acres a year into perspective, we can estimate that each year an area of forest larger than the state of California disappears from the earth.

What does the loss of these trees, insects, and plants have to do with us? Let's start with the trees. In their growth process, trees both store carbon dioxide and convert it into living tissue—wood. In one day a single deciduous tree can absorb about three-quarters of an ounce of carbon dioxide from the atmosphere, roughly 16 pounds a year. When that tree is burned, it not only releases carbon dioxide, it removes one of nature's devices for absorbing it.

As a result, the carbon dioxide released through the burning of tropical forests is thought to contribute at least 20 percent of the global warming problem. (As stated earlier, carbon dioxide is responsible for about 50 percent of the greenhouse effect.) Also released during burning are methane and nitrous oxides, which contribute further to the warming problem.

Of course, not only trees are lost during the burning and logging. Conservative estimates put the number of plant, animal, and insect species native to rain forests at two million. Most estimates put the number somewhere between 5 million and 80 million species; one researcher has estimated that there are as many as 30 million insect species in tropical forests alone. Although only a fraction of 1 percent of these species vanishes each year—approximately 4,000 to 6,000 species are lost to deforestation annually—the rate is thousands of times greater than the natural rate of extinction. Indications are that, as deforestation continues, the rate of extinctions could become higher still.

As with the trees, these insects and plants affect our daily lives, and their demise could have immediate economic consequences. In 1970, for example, a leaf fungus blighted U.S. cornfields from the Great Lakes to the Gulf of Mexico, eliminating 15 percent of the entire crop and pushing up corn prices by 20 percent. Losses to farmers exceeded $2 billion. The situation was saved by a blight-resistant germ plasm whose genetic ancestry traces back to variants of corn from one of the plant's native habitats in Mexico.

Moreover, according to Norman Myers, a consultant on the environment and development, half of the purchases in your neighborhood pharmacy, whether drugs or pharmaceuticals, derives from wild organisms. The full commercial value of these wildlife-based products worldwide, says Myers, is $40 billion a year. Remember those poison darts used as weapons by tribesmen in Tarzan movies? That "poison," curare, derived from a South American tree bark, is an important anesthetic used during heart, eye, and abdominal surgery. And curare can't be synthesized in the laboratory. Derivatives from the rosy periwinkle offer a 99 percent chance of remission for victims of lymphocytic leukemia, as well as a 58 percent chance of recovery from Hodgkin's disease.

There are thousands of tropical plants that have become valuable for industrial use. First and foremost is rubber. Today, over half the world's commercial rubber is produced in Malaysia and Indonesia. The sap from Amazonian copaiba trees, poured straight into a fuel tank, can power a truck; 20 percent of Brazil's diesel fuel comes from that tree.

And then there is food. Thousands of tropical plants have edible parts, and some are superior to our most common crops. In New Guinea, for example, the winged bean, *Psophocarpus tetragonolobus*, has been called a one-species supermarket: the entire plant—roots, seeds, stems, leaves, and flowers—is edible, and a coffeelike beverage can be made from its juice. It grows like a weed, reaching a height of 15 feet in a few weeks, and has a nutritional value roughly equivalent to soybeans. There are thought to be many such plants that may be developed for commercial use in this and many other countries, creating foods to feed a hungry world.

All told, humans have exploited less than one-tenth of 1 percent of naturally occurring species. And as thousands disappear each year, along with them go their potential.

Agriculture in tropical rain forests is possible, say the experts, if practiced according to the principles of "sustainable use." This refers to any ongoing activity that doesn't permanently degrade the forest, such as the collection of rubber, nuts, or herbs. Environmentalists working with the timber industry have attempted sustainable logging programs, but such plans haven't yet been put into large-scale use.

**What You Can Do.** It may seem unfeasible to directly affect the burning and harvesting of tropical forests thousands of miles away, but there are actions you can take. Among them:

❑ **Don't buy wood harvested from endangered tropical forests.** Instead, look for woods that are forested in temperate regions of Europe, North America, and Japan, or are harvested in sustainably managed tropical timber projects. These include ash, beech, birch, cherry, elm, hickory, oak, poplar, and black walnut.

❑ **Avoid buying beef products from livestock raised in tropical regions.** Indentifying such beef, however, is not always easy.

Some rain forest beef is labeled "American beef."

❑ **Support sustainably managed tropical products.** Examples are nuts and cosmetics made from rain forest herbs. Most such products provide this information on their labels; some are mentioned in the "Food and Groceries" and "Personal Care Products" sections of this book.

## 6. Garbage

The newspaper and TV news stories a few years back about the trash-filled barge with no place to dock didn't even begin to describe the extent of the garbage crisis we are facing. Some cities are virtually choking on their citizens' refuse, and solutions are few and far between. The simple fact is that Americans generate so much garbage that we are rapidly running out of places to put it.

Although the problem has been around for years, the problem seems to be getting worse. Between 1960 and 1986, the amount of American garbage discarded annually grew by 80 percent, to between one and two tons of melon rinds, grass clippings, plastic hamburger boxes, and discarded toasters for every living soul in the country. All told, Americans throw away an average 3.5 pounds of solid waste every day, although some estimates range as high as 6.5 pounds a day. By comparison, the average West German or Japanese throws away about half as much.

The obvious problem is where to put it all. Sanitary landfills—what we used to call "dumps"—are rapidly filling up, sometimes turning away trash generated by those in other regions, as the barge-to-nowhere episode so graphically illustrated. There's good reason for this glut of garbage: 80 percent of all our trash goes into landfills; of the remaining 20 percent, half is incinerated and half is recycled. The U.S. Conference of Mayors predicts that over half of the nation's 9,300 landfills will face closure within ten years.

But where to put the trash is only part of the problem. An equally serious question has to do with what is contained in our trash. Experts believe that somewhere between 5 percent and 15 percent of municipal solid waste contains hazardous substances—garbage that can injure living things and is sometimes even life-threatening. Such hazardous wastes go well beyond the leaky drums of industrial chemicals we've seen on television. They include many of our most common household items.

Another problem with our garbage is that so much of it is made of plastic and other materials that do not break down easily or quickly. Indeed, we do not even know how long plastic, polystyrene, and similar common substances will last, given that such materials have been invented only in the past few decades. Best guesses for polyethylene, for example, the substance from which disposable diapers are made, is that it will break down in three hundred to five hundred years. But this is entirely speculation; no one really knows. (See "Diapers, Diapers, and More Diapers" for more on this subject.)

What exactly is in our trash? According to a 1988 study by Franklin Associates of municipal solid waste (which includes primarily household trash, as opposed to commercial, institutional, and industrial sources), here is the breakdown, listed by percentage of the total volume:

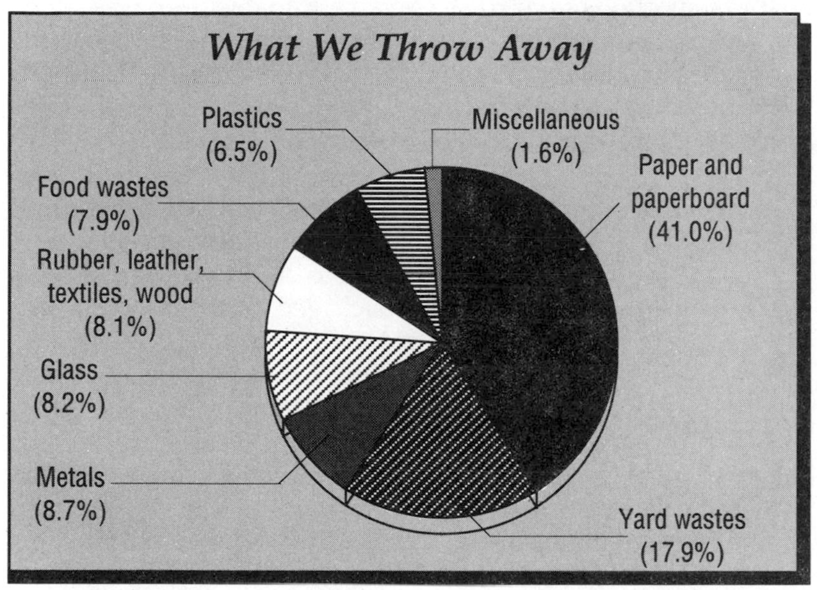


This list doesn't exactly seem like a toxic chemical dump. Or does it? Here's a sampling of some of what's in our daily trash:

❏ **Batteries** — used in flashlights, radios, cordless appliances, and dozens of other things—contain cadmium, lead, lithium, manganese dioxide, mercury, nickel, silver, and zinc, all of which are toxic to humans, and all of which can leak out of corroded batteries when they are dumped in landfills.

❑ **Glass containers, tableware, and cookware** contain lead, which, when it leaks into the water supply, can severely impair children's mental and physical development.

❑ **Plastics** contain polyvinyl propylene, phenol, ethylene, polystyrene, and benzene, all of which are among the most hazardous air pollutants around.

❑ **Used disposable diapers** can contain any of over 100 viruses, including live polio and hepatitis from vaccine residues contained in feces.

The tragic fact behind such hazards is that most of what we throw into landfills can be recycled, turned into compost, or otherwise disposed of safely. Recycling in particular makes sense. A sizable portion of our trash contains valuable resources—metals, glass, paper, and wood—that can be reprocessed and used again. Cities and states throughout the United States are adopting voluntary or mandatory recycling programs, usually with great success. (More on this in "The Recycling Solution.")

**What You Can Do.** There's no doubt that recycling is the single most important step in easing the garbage crisis. But even before that comes *pre*cycling—your decision to avoid buying materials that are overpackaged or that are packaged in harmful plastics or other materials. (See box, page 42.)

## 7. Water Pollution

We've come to expect safe drinking water in our homes. Turn on the tap, fill a glass, drink it up, don't give it a second thought. But things just aren't that simple. Despite the years of concern about water pollution, Americans still dump 16 tons of sewage into their waters—every minute of every day. Unfortunately, many of the environmental problems mentioned above—air pollution, acid rain, garbage—end up in our drinking water supply. And even with bottled water, which many people drink to avoid these pollutants, there are environmental problems.

Fresh water is one of the world's most precious resources. Despite the fact that two-thirds of our planet is covered with water, all the fresh water in the world's lakes, creeks, streams, and rivers represents less than .01 percent of the earth's total water. Fortunately, this freshwater supply is constantly replenished through rain and snow. Unfortunately, much of the rain and snow is contaminated by gases and particles that human activity has introduced into the atmosphere.

When we think of water pollutants, we typically picture huge industrial plants spewing millions of gallons of muck into our rivers and streams. While industry does account for a portion of water pollution, treatment of industrial wastes has improved considerably over the past decade. Only about 9 percent of water pollution today comes from direct industrial dumping into rivers, streams, and oceans. Today, concern has been focused on an equally serious problem: groundwater pollution.

If you dig deep enough at almost any spot in the United States, you're likely to hit water. The volume of groundwater within a half mile of the U.S. land surface is said to be four times that of the Great Lakes. Nearly half of all Americans and three-fourths of the urban dwellers rely on groundwater as their primary source of drinking water, using an estimated 100 billion gallons a day.

Underground water picks up whatever it passes through. Rainwater and melted snow—running off parking lots, rooftops, streets, and farms—carry with them deadly substances from worn brake linings, chemical fertilizers, old tires, and a variety of other materials contained in our garbage. During a storm, the pollutants are washed into streams and rivers. Once they get into the water cycle, they never seem to leave.

There are other sources of pollution:

❑ **Underground gasoline storage tanks**, most of them located at gas stations, leak petroleum products into the soil. The federal government has estimated that there may be as many as 100,000 leaky gasoline storage tanks in the United States.

❑ **Agricultural irrigation**, which accounts for about two-thirds of all water use, can leach salts and pesticides from the soil and into the water. The Environmental Protection Agency has estimated that each year between 3.5 million and 21 million pounds of pesticides reach groundwater or surface water before degrading.

❑ **De-icing salts**, used throughout the country's frost belt, have caused contamination in eleven northern states as they dissolve and percolate into the soil.

One possible solution might be to drink bottled water, assuming that everyone could afford that luxury. But most bottled water, despite advertising and marketing campaigns that might suggest otherwise, isn't necessarily more healthy than most cities' tap water; it just tastes better. The regulations currently enforced by the Food and Drug Administration set standards for bottled water that are very similar to the EPA's standards for tap water. And the tap water standards are not known for being strict. The rules set "tolerance" levels for some very potent chemicals, but they don't specify any limits for a wide variety of synthetic compounds called "organics"—an alphabet soup of carcinogens and mutagens such as polyvinyl chloride, trihalomethanes, and polybrominated biphenyls (PBBs). Around 700 organics have been measured in drinking water so far; many of them have never been tested for toxicity.

In addition, bottled water, however tasty, encourages a wasteful and polluting industry. For one thing, the energy involved in moving water around the world in bottles is appalling. In addition to transportation fuel, there is the cost of packaging. It takes the energy equivalent of a ton of coal to produce a ton of glass; plastic bottles, of course, contain other notorious pollutants; about a third of bottled water is sold in nonbiodegradable and nonrecyclable plastic containers. Moreover, the plastic polyethylene containers are known to leach their chemicals into the water, which mixed with other impurities, can be toxic. Another solution—investing in a household water-filtering device that is connected to your tap—may be effective (although some filters create health hazards themselves; see "Fresh Drinking Water" for details), but it can be costly. The best solution is to work toward improving the existing water supply in your community. (See "How to Get Involved" for more on this.)

Beyond the issue of water pollution is the question of water supply: Is there enough water available for everyone? The good news is that there is plenty of fresh water on the earth—about 9,000 cubic kilometers—which, according to one estimate, is enough to sustain 20 billion people, more than four times the earth's population. The bad news is that the water and its people are unevenly distributed. So some areas (like Iceland) have far more water than they can possibly use; many other areas have far less than they need.

In the United States, there are fewer extremes. While some areas face water deficits, there are relatively few areas with excess water supply. An increasing number of communities are turning to water conservation as a means of making scarce water resources available to everyone or of ensuring that adequate water supplies remain reliable.

Besides ensuring an adequate supply, there is at least one additional benefit to conserving water. Water conservation helps protect fish and wildlife habitats and wetlands. These environments may be irreparably damaged if there is not enough water to sustain them, with serious ecological consequences and, in some communities, economic impacts as well.

By diverting less water, we leave more water for other uses, from boating and fishing to power generation. Maintaining adequate flow also helps ensure water quality: water is more susceptible to contamination when there is a low flow. And by using less water we also discharge less used water, thereby further reducing the potential for pollution. Finally, using less water minimizes the need to build new dams and reservoirs, costly projects that can further destroy wildlife.

**What You Can Do.** There are two areas of concern: conserving water and keeping it clean.

❑ **Don't waste water**. Much of the water we use at home is wasted. From installing water-saving devices to making changes in your daily habits, there are many ways to conserve water without compromising convenience.

❑ **Don't add pollutants at home**. Some toxic products—such as bleaches and paints—should not be dumped directly down the drain. There are other environmentally safe methods of disposal.

❑ **Buy products that don't pollute the water**. Some soaps and detergents, for example, contain phosphates and other substances that can harm the water supply. Fortunately, there are safe alternatives to most of these products.

❑ **Report water polluters**. If you believe a factory or other business in your area is dumping harmful substances into local lakes or streams, report them to local or federal officials; additional information is contained in Part III of this book.

## The Problem of Packaging

We have become the most overpackaged society in history.

We love packaging. And the products in our supermarkets, drugstores, and hardware stores reflect that love affair: nearly everything, it seems, is wrapped in something. Even produce—tomatoes and corn on the cob, for example—sit neatly on a plastic foam tray, encased in clear plastic wrap. Some products have layers upon layers of wrapping, for no apparent practical reason.

Guess who's paying for it all? You are, in a number of ways. Packaging not only adds to the price of these products, it also costs you money to dispose of it or to repair the harmful effects many of the packaging materials have on the environment.

Make no mistake: packaging is important. It protects products from damage, ensures that they remain sanitary, prevents tampering, provides space for product information, and offers convenience (can you imagine carrying home all those soft drinks if they didn't come in six-packs?). But the manufacture, use, and disposal of packaging materials contribute to many environmental problems, from litter to acid rain.

We've already discussed the solid waste problem (see "Seven Environmental Problems You Can Do Something About"). But how much do packaging materials contribute to the problem? Of the roughly two tons of trash discarded by the average American each year, packaging accounts for an estimated 30 percent—about 1,200 pounds of packaging trash a year for every man, woman, and child. For the nation, those statistics mount up into some pretty impressive numbers. For example, according to the Environmental Defense Fund:

- ❑ We throw away enough glass bottles and jars to fill the 1,350-foot twin towers of New York's World Trade Center *every two weeks.*
- ❑ We toss out enough aluminum to rebuild our entire commercial air fleet *every three months.*
- ❑ We go through 2.5 million plastic bottles *every hour.*
- ❑ We discard enough iron and steel to *continuously* supply all the nation's auto makers.

As the major food manufacturers vie for highly competitive

## *The E-Factor*

What do you look for when you shop? If you are a typical consumer, you probably consider three things about a product before deciding: price, quality, and convenience. To be a Green Consumer, you should add a fourth: the E-factor.

"E," in this case, stands for "environment," but it also stands for "ethics." However you define the term, an increasing number of consumers are examining these additional qualities of products they buy, from toothpaste to tires to televisions. How green is the product? How green is the company behind the product? In addition, how socially responsible is the company—what role does it play in equal opportunity, national security, minority advancement, and military contracts, among other important issues?

These are not always easy decisions to make. For one thing, answers often are largely subjective, requiring you to impose your own values on these companies. If a company makes an otherwise good product, does it matter to you how many women and minorities it hires or whether it contributes to the arts? Maybe. Maybe not. It ultimately is a personal decision.

In addition, as stated earlier, things usually are not simply black or white. There are polluting, socially irresponsible companies making green products, and well-meaning companies making products of marginal quality, value, or effectiveness. It is difficult at times to keep track of all the various issues and concerns.

One excellent resource for this information is *Shopping for a Better World*, subtitled "A Quick & Easy Guide to Socially Responsible Supermarket Shopping," published by the **Council on Economic Priorities** (CEP, 30 Irving Pl., New York, NY 10003; 212-420-1133). The 140-page guide rates the positions of 1,300 brand-name products on eleven key social issues, from minority advancement to military contracts, charitable contributions to community outreach. To order *Shopping for a Better World*, send $5.95 (including shipping) to CEP at the address above. The book, published by Ballantine, is also available in bookstores. CEP also offers "Shopping for a Better World: The List" ($5, including shipping), a pad of shopping lists that include names of products made by companies that received 8 or more top ratings in *Shopping for a Better World.*

supermarket shelf space, many have created innovative packaging techniques that enable their products to stand out in the crowded marketplace. But in their attempts at innovation, many of these manufacturers are creating wasteful products, with prices to match. Consider, for example, Souper Combo, introduced in 1988 by the Campbell Soup Company. The product, a soup-and-sandwich heat-and-serve convenience meal, consists of *six separate layers* of packaging, five of them plastic: a polypropylene bowl, two polystyrene trays, a polyester soup bowl lid, a polyester sandwich wrap, and a container made of paperboard. Only the paperboard is recyclable. All of this to encase 11.9 ounces of food. *Food Business Magazine* called Souper Combo one of the five best products of 1988; *Food and Drug Packaging Magazine* named it "Product of the Year."

However much admired, such products leave a lot to be desired, environmentally. They represent a wasteful use of resources and a lot of extra trash. Moreover, this trash consists primarily of plastics, which can contribute toxic substances to soil and water. Even worse, these plastics are likely to remain intact for centuries.

## The Perils of Plastic

There is no question that plastic is the packaging material of choice for most product manufacturers. Since the 1940s, manufacturers have come to depend on a seemingly endless variety of plastic packaging materials: bottles, bags, jars, jugs, tubes, tubs, foams, films, pellets, wraps, and so on. Today, the volume of plastics used in the United States exceeds that of steel. The plastics industry predicts that consumption of plastic resins will grow an estimated 50 percent during the 1990s. The single largest use of plastics today is packaging, constituting a fourth of the 12 billion or so pounds of plastics produced each year in this country.

The debate over plastic is an intense one. Both environmentalists and the plastics industry generate endless reams of paper—recycled, we hope—on the subject, and each side has well-crafted arguments. Indeed, if you were to listen only to the plastics industry, you would be led to believe that its miracle material has helped to save the environment. One industry group, for example, provides reams of data "proving" that, compared to glass and

## *Baggies from Plants (and Spuds)*

Every day, millions of brown-baggers concoct a mouthwatering sandwich at home, slip it into a plastic sandwich bag, and carry it to work or school. At lunch, the sandwich is eaten and the bag is tossed into the trash. A single plastic sandwich bag won't cause a major landfill problem, but one plastic sandwich bag a day multiplied by millions of employees and students can produce a mountain of nonbiodegradable plastic trash.

This doesn't mean we have to give up those convenient, perfectly sized protective bags. New on the market are cellulose food bags. Made from a plant fiber used in paper making since the second century, the bags are nontoxic and estimated to biodegrade in one to three years. Of course, the best solution is to pack lunch in a reusable container, but for those who must have sandwich bags, cellulose is the next best thing. Prices are higher than those of traditional plastic bags, usually ranging from about $4 to $6 for 50 sandwich-size bags, and $6 to $8 for 50 larger, "freezer-size" bags. Suppliers of cellulose bags include:

Seventh Generation
10 Farrell St.
South Burlington, VT 05403
802-862-2999, 800-456-1177

Earth Care Paper Co.
100 S. Baldwin
Madison, WI 53703
608-256-5522

Co-Op America
2100 M St. NW, Suite 310
Washington, DC 20036
202-872-5307, 802-658-5507

Ecco Bella
125 Pompton Plains Crossroads
Wayne, NJ 07470
201-890-7077

Cellulose is only one alternative material suitable for bags. Another is potato peelings. According to Argonne scientists, an estimated 10 billion pounds of potato waste are created each year from the peeling and cutting of potatoes for french fries alone. In 1989, scientists at the Argonne National Laboratory in Illinois announced that they had developed a potato-based plastic that biodegrades faster than most other products, including cellulose. The plastic sheets, which can be made into garbage and grocery bags, decay not only in the presence of sunlight, but also in water. Moreover, they are edible to bacteria.

aluminum, PET, the plastic used for beverage bottles, is more energy efficient, creates less pollution, and is fully recyclable. Environmentalists, of course, have a much different point of view.

As Jeanne Wirka, coordinator of the Solid Waste Alternatives Project at **Environmental Action**, points out, the use of plastics has a two-pronged effect on the environment:

❑ Many of the chemicals used in the production and processing of plastics are highly toxic. Most of the plants that produce these chemicals also produce hazardous wastes that pollute air and water. In 1986, when the Environmental Protection Agency ranked the 20 chemicals whose production generates the most hazardous waste, five of the top six were chemicals commonly used by the plastics industry: propylene, phenol, ethylene, polystyrene, and benzene.

❑ When the relatively short lifetime of these plastic products (a foam coffee cup or fast-food hamburger container may be in use for scarcely a few minutes) is over, this trash frequently ends up littering streets, parks, rivers, and oceans. And, of course, it contributes to our solid waste problem. As the U.S. General Accounting Office, the investigative arm of Congress, put it: "Some packaging used for fast foods has an estimated service life of only three minutes but may persist as litter for centuries."

The detrimental effects of plastics aren't just in the air and water. They also affect many living creatures. The world's oceans, for example, are clogged with floating plastic debris. One marine study group concluded that the garbage floating in the world's seas now outweighs the fish harvest by a factor of three to one. Plastic six-pack yokes, Styrofoam pellets, plastic bags, and tampon applicators (which have become derisively known as "New Jersey seashells") are among the most common materials bobbing in the waves. The thousands of fish, birds, and marine mammal species that feed off the water surface cannot distinguish these things from food. Indeed, much of it resembles their very diet: floating plastic bags are mistaken for jellyfish; bits of Styrofoam resemble fish eggs; cigarette butts look like small brown crustaceans. The results are tragic, with wildlife choking, starving, or developing lethal infections from this sea of trash. A report from the federal Office of Technology Assessment concluded that plastic pollution is a greater threat to marine mammals and birds than are pesticides, oil spills, or contaminated runoff from land.

## Burn, Bury, or Recycle?

Getting rid of plastics isn't easy. There are basically three choices: burning it, burying it, or recycling it. Let's take a brief look at each.

❑ **Burning** plastics in incinerators is a risky business. Most plastics release a host of toxic materials when burned, including such heavy metals as cadmium, lead, and nickel. These substances are emitted into the air or are contained in the ashes left behind—which still must be disposed of somehow.

❑ **Burying** plastics, of course, has its own problems, not the least of which is the shrinking amount of landfill available to hold a growing amount of plastic trash. But a growing amount of plastic is being labeled as "biodegradable" or "photodegradable"—that is, it is designed to disintegrate or decompose over time. Materials biodegrade when micro-organisms absorb or eat portions of a material, causing it to break down into smaller pieces. Other materials photodegrade due to a chemical reaction initiated by direct exposure to ultraviolet light.

But researchers have found that plastics—and many other materials, including newspaper—can take years to break down once they are buried in landfills. There are many stories of perfectly readable twenty- or thirty-year-old newspapers uncovered from the bottom of landfills; in effect, they aren't so much discarded as mummified. The reason: For biodegrading or photodegrading to take place, oxygen or sunlight, respectively, must be present. At the bottom of a landfill, they aren't.

Some plastics product manufacturers have taken to adding starches or cellulose to their plastics mixtures, which is intended to speed up the degrading process. Both of these substances are more readily eaten by microbes, and thereby the degrading process is accelerated. While such products—generally marketed as "biodegradable"—are preferable to those that don't contain them, the addition of starch or cellulose doesn't necessarily solve the problem of what happens to these plastics at the bottom of a dark, airless landfill. Evidence shows that they, too, can remain intact for years.

Even when plastics do biodegrade or photodegrade, questions remain. For instance, what materials do plastics disintegrate into? The liquid that filters through degradable waste, called "leachate," picks up toxic substances, such as heavy metals used as pigments and stabilizers in plastics, which can subsequently contaminate groundwater. The gaseous by-product of the biodegradation process, methane, is explosive and unhealthy to breathe. Another problem has to do with the by-products formed during the degradation process. For example, during degradation small pieces of plastic may be created. There is a concern that these small pieces may pose risks to birds and other wildlife that mistake them for food.

The bottom line is that biodegradable plastics, at most a short step above their nonbiodegradable versions, are still far from desirable. They do not solve the problem of shrinking landfill space, and they may create new contaminants to the environment. Even the federal government, usually supportive of industry efforts to solve environmental problems, is skeptical. As Jonathan Z. Cannon, an official with the Environmental Protection Agency, stated in 1989: "Unfortunately, at this time, we cannot be sure that the use of biodegradable plastics will prove superior to nonbiodegradable plastics."

❑ **Recycling.** That leaves the third option, recycling, one that is being promoted aggressively by the plastics industry. As the Plastics Recycling Foundation, an industry group, put it: "We fervently believe that plastics hold the potential for becoming the most recyclable packaging material."

But recycling plastic is far different from recycling glass, aluminum, cardboard, or paper. While the recycling process for these materials is continuous—you can recycle an aluminum can or glass bottle endlessly—most plastic recycling is a one-time thing. Researchers have developed processes to shred plastics—primarily from soda bottles—into flakes, which are then washed and separated from residual materials. The plastics recycling industry has found many ingenious uses for this shredded plastic: furniture cushions, bathtubs, flowerpots, pipes, toys, trash cans, shipping pallets, parking lot car stops, boat docks, and insulation, among a growing number of other products. But these are mostly one-time uses. All of these products have limited lifetimes, and they, too, will need to be disposed of eventually. It is unclear whether there will be any technology to take these parking lot car stops, flowerpots, and other items to reuse their plastic once again.

In an attempt to encourage recycling of plastics, the Plastic Bottle Institute, a trade organization, established a voluntary material identification system of plastics recyclability. The system was created primarily to assist materials separators in identifying which plastic bottles could be recycled. If you must purchase beverages in plastic bottles, this system can help you also to choose bottles that have the potential for recycling. Please understand: *This system does not ensure that your plastic purchase will be recycled. Therefore, it does not make plastic an environmentally acceptable packaging material.*

The system features seven codes, one of which is usually stamped on the bottom of a bottle. Please note that only three of the plastics types are easily recyclable. Here is a brief summary of what they mean.

1. **Polyethylene terephthalate (PET)**—23 percent of all plastic bottles. This type is used to package boil-in-bag foods, meat, cosmetics, and carbonated soft drinks. When recycled, these containers are turned into a wide range of fibers, textiles, fiberfills, polyester, and engineering plastics, to name just a few of the applications.
2. **High-density polyethylene (HDPE)**—62 percent of all plastic bottles. This is the most common type of plastic used for consumer products, including milk bottles, liquid detergents, shampoos, pharmaceutical products, juices, bottled water, and antifreeze. When recycled, HPDE is used for plastic fencing, garden furniture, base cups for soft drink bottles, flowerpots, drainage pipes, toys, milk bottle crates, and many other things.
3. **Polyvinyl chloride (PVC)**—6 percent of all plastic bottles. These are used to package floor polishers, shampoos, edible oils, mouthwashes, and liquor. When recycled, the plastic is used for drainage and sewer pipes, vinyl floor tiles, truck-bed liners, and down spouts, among other things.
4. **Low-density polyethylene (LDPE)**—4 percent of all plastic bottles. This lightweight, squeezable material is used in packaging toiletries, cosmetics, and food products. *Low-density polyethylene is not easily recyclable.*

**5. Polypropylene (PP)** —4 percent of all plastic bottles. Because of its high heat resistance, this is used to bottle "hot-filled" foods (products that must be put in containers while hot) such as syrup. *Polypropylene is not easily recyclable.*
**6. Polystyrene (PS)**—1 percent of all bottles. This is used to package tablets, salves, ointments, and other products not sensitive to oxygen and moisture. *Polystyrene is not easily recyclable.*
**7. Other plastics.** This includes a wide range of substances, including all other resins and packaging that include layers of plastic and other materials. *None of these plastics is known to be easily recyclable.*

Perhaps the biggest problem associated with this coding system is that there are few household trash recycling centers that accept plastic bottles. Obviously, this makes recycling plastic even more difficult and decreases the chance that the recyclable plastic bottle or container you buy will ever be recycled. Still, despite the system's shortcomings, the possibility that your empty container will be reused even just once is better than if it is never reused. So, plastic recycling does play a role. But it probably will not be the saving grace that the plastics industry seems to believe it will be.

## Refuse, Reuse, and Recycle

What, then, is the solution to the plastics problem? As with so many other environmental problems, the solution requires a varied, integrated approach. There are three key parts:

❑ **Refuse to buy plastic.** The perfect solution is to avoid buying anything made of or packaged in plastic. While a total avoidance of plastic is more or less impossible unless you live on a self-sustaining farm, you can avoid it wherever possible. Don't buy produce wrapped in plastic. Ask for paper instead of plastic bags at the supermarket. (See "Paper Bags or Plastic?" for more on this.) Buy clothing made of natural instead of synthetic materials. Opt for glass, aluminum, or cardboard packaging. Avoid fast-food or carry-out restaurants that package their food in foam or plastic. (There are many more solutions throughout Part II of this book.)

❑ **Reuse plastic products.** Whenever possible, reuse plastic containers—for storing food, hardware, or other small items. Also, look for products in containers that can be refilled without your buying an additional plastic container.

❑ **Recycle plastic products.** If you must buy plastic, seek out those that can be recycled. Keep these plastics separate from other trash, then bring them to a recycling center that accepts plastics.

As you can see, there is no simple solution to the perils of plastics. For now, plastic packaging is a reality in our lives; it is almost impossible to avoid it completely. Still, the less plastic packaging you buy, the better. The best solutions for Green Consumers are those that use—and reuse—plastics in a responsible way.

## THE RECYCLING SOLUTION

Let's start with the basics: Recycling makes sense. It is both economical and ecological. For Green Consumers, no activity has a more direct payoff to the environment. Problems from water and air pollution to landfill overflow and the greenhouse effect—in other words, just about every environmental problem faced in the 1990s—can be eased through recycling.

Recycling is not a new idea. Far from it. Half a century ago, during World War II, recycling was a way of life for many Americans. Everything from tin cans to scrap iron, from rubber to cooking grease, was recycled to help the war effort. Everyone did his or her part to preserve the country's scarce or strategic resources. When the war ended and manufacturing boomed, Americans developed a throwaway mentality. Besides, America is such a large country that no one figured we would run out of space to bury our trash.

Now, reality has set in, and recycling has come full cycle. America's scarce and strategic resources may have changed, but the solution to preserving them has remained the same: recycling.

Consider just a handful of the benefits to the environment recycling can bring:

❑ Using recycled paper instead of virgin paper for one print run of the Sunday edition of *The New York Times* would save 75,000 trees.

## *The New Art of Precycling*

In the next chapter, we'll closely examine the world of recycling, and the financial and environmental benefits gained by giving a wide range of materials and resources a second life. For now, however, let's consider a completely different and somewhat revolutionary thought:

**You can avoid dealing with much of this waste by precycling.**

That's right: *pre*cycling. Precycling means making intelligent, environmentally sound decisions at the store, and reducing waste *before* you buy. In other words, by precycling you will reduce waste by avoiding buying those things that will need to be disposed of or recycled.

In 1989, the city of Berkeley, California, launched an innovative citywide precycling campaign, complete with posters, brochures, and newspaper ads. Other communities and organizations have examined this relatively simple notion—which is known more formally as "source reduction." By whatever name, in whatever form, the idea makes good environmental sense.

Clearly, you won't be able to precycle all your products. There simply aren't reasonable alternatives for some purchases. But the more you do precycle, the less trash you will generate.

How do you precycle? Here are six basic guidelines:

❑ The amount of copper thrown away each year—primarily from electronics—in a city the size of San Francisco is more than the amount produced annually by a medium-size copper mine.
❑ Recycling creates six times as many jobs as does landfilling and incineration.
❑ A steel mill can reduce its water pollution by 76 percent and its mining wastes by 97 percent using recycled scrap metal instead of iron ore. A paper mill can reduce its air pollution by 74 percent using recycled waste paper instead of virgin pulp.
❑ In 1988 alone, recycling of aluminum cans saved more than 11 billion kilowatt-hours of electricity, roughly enough to supply the residential electric needs of New York City for six months.

1. **Be picky about packaging.** At the store, consider more than just the contents of a package; consider the package itself. Can it be reused, refilled, or recycled? Whenever possible, choose products packaged in recyclable materials such as paper, cardboard, glass, and aluminum.
2. **Don't pay for overpackaging.** If the packaging isn't necessary to protect the product, buy the less-packaged alternative. There's a good chance it will cost less, too.
3. **Avoid plastic.** Even the plastic that is recyclable has a limited life compared with glass, aluminum, and cardboard. Eventually, even recycled plastic products will contribute to environmental problems. Be particularly wary of polystyrene foam (Styrofoam is one brand), which can be harmful to the stratospheric ozone layer.
4. **Don't buy disposables.** The longer the life span of a product, the better. Disposable razors, diapers, lighters, cameras, cups, and plates are a waste of resources and contribute to disposal problems.
5. **Buy in bulk.** Whenever practical, purchase larger packages. You'll save money and avoid excess containers. Especially avoid "convenience" packaging—single-serving meals, for example—which are both overpackaged and overpriced.
6. **Look for products in refillable containers.** A growing number of products include reusable containers. From cleaners to cake mixes, manufacturers are creating new products that require you to purchase the container during the first purchase, then buy only refills or concentrates after that, eliminating the need to purchase unneeded packaging.

These impressive statistics notwithstanding, Americans currently recycle only about 10 percent of their household trash. In the process, they throw away incalculable amounts of raw materials—and the energy it takes to convert these materials into finished goods. Part of the problem may be that many people don't understand what can and what cannot be recycled. According to a poll by the Gallup Organization, Inc., 54 percent of Americans said that if they knew a certain food or beverage container was not recyclable, they would switch to a container that was.

There are at present some 8,000 community recycling programs in the United States. These range from voluntary drop-off programs, in which individuals bring recyclable materials to central locations, to mandatory curbside pickup programs, in which citizens are required by law to separate their garbage and prepare it in a prescribed manner for pickup by the local trash company. In addition, a growing number of business recycling programs are emerging, including some mandated by state or local law, in which companies save, reuse, and recycle the massive amounts of paper, cardboard, metals, and other materials used in our information and manufacturing society.

Recycling is really only part of what has become known as "integrated waste management"—a bureaucratic term that refers to a four-part strategy for effectively dealing with landfill and pollution problems: *reducing* the amount of trash generated; *incinerating* refuse, preferably turning the heat into energy; *recycling* as much as possible; and *landfilling* whatever is left. The federal government has actively been promoting these integrated programs, which can be made flexible enough to meet each community's needs.

## What's Recyclable?

Not everything we buy can be recycled, but most things can. A study conducted on Long Island by the Center for the Biology of Natural Systems found that 84 percent of household waste—including food, yard waste, paper, bottles, and cans—can be recycled. Of course, the exact percentage of recyclable materials you throw away is directly related to how much you consider recyclability when shopping. (See "The New Art of Precycling" for more on making smart choices.)

What exactly is recyclable? Here, in alphabetical order, is a status report on the recyclability of our most common products and materials.

**Aluminum.** Aluminum is one of the most expensive and polluting metals to produce. For one thing, the metal is extracted from bauxite ore, which is mined on the surface, much of it in tropical rain forest areas. Fortunately, recycling aluminum cans is one of the great environmental success stories of recent years. Since the early 1970s, can recycling has grown steadily; Americans now recycle about 55 percent of all aluminum cans—some 42 billion recycled cans containing about 1.5 billion pounds of aluminum. According to the Aluminum Association, that represents more than 5 million tons of aluminum that has not been buried in our landfills since 1972.

The aluminum recycling process is relatively swift: Used aluminum cans are melted down and returned to store shelves in the form of new beverage containers in as few as six weeks. One reason for such speed is the ease of recycling an all-aluminum can: Because most aluminum cans have no labels, caps, or tops, they needn't be separated from other "foreign" materials before being reused.

Nearly all recycling facilities accept all-aluminum cans. (To tell the difference between an aluminum can and a steel one, apply a magnet. If the magnet sticks, the can is steel.) Some centers accept bimetal cans—those with aluminum top and steel sides. In addition to cans, you may include aluminum foil, pie pans, and other aluminum scrap. You needn't prepare aluminum for recycling except to separate it from other trash. For your convenience, however, it might help to crush cans before discarding. Of course, in many areas, you can return the cans to stores for a refund.

**Batteries.** The number of dry-cell batteries we use is growing by nearly 10 percent a year. We now throw away about 2.5 billion batteries—from the cylindrical flashlight variety to the tiny button-size type used in cameras, watches, calculators, and hearing aids. These innocent-looking batteries are rarely considered environmental threats, but they contain some very toxic chemicals. When burned in incinerators, the heavy metals in batteries—including cadmium, lead, lithium, manganese dioxide, mercury, nickel, silver, and zinc—pollute the air or become toxic components of discarded incinerator ash. When tossed into landfills, the metals leach out of corroded batteries and seep into the groundwater. These metals are so dangerous that the Occupational Safety and Health Administration has established workplace exposure limits for all eight metals mentioned above.

Battery recycling is done routinely in Europe and Japan, but it has only begun to catch on in the United States, which has only a handful of battery recycling programs. One U.S. company, **Mercury Refining Company** (790 Watervilet Shaker Rd., Latham, NY 12110; 518-785-1703, 800-833-3505), has been reclaiming the mercury and silver from batteries since the 1950s. The company accepts all types of household batteries. It separates the various types for different treatment: alkaline and carbon-zinc batteries go in hazardous-waste landfills. Mercury, nickel-cadmium, and silver batteries are sent to another facility for recycling. Mercury Refining Company personnel are available to assist local community groups in setting up battery recycling programs.

Another solution to dry-cell battery pollution is rechargeable batteries. Although these batteries cost more and require the one-time added expense of a recharger, most such batteries can be recharged up to one thousand times. Then, the nickel and cadmium can still be reclaimed and recycled.

**Automobile batteries** represent an even more insidious pollution problem. Each year, about 80 million lead-acid car batteries are discarded. Although service stations used to pay a few dollars for dead batteries, now some won't accept them at all. Most batteries wind up in landfills, where their lead can seep into groundwater. (Lead is a highly toxic substance, which causes anemia, nerve damage, and paralysis; about two-thirds of the lead consumed in the United States is used for lead-acid batteries.) In addition, the sulfuric acid contained inside the batteries leaks out, dissolving almost everything it touches before it too seeps into groundwater. A few jurisdictions are beginning to require deposits on car batteries as a means of keeping them out of the local waste stream. In Suffolk County, New York, for example, a $5 refundable deposit has proved effective.

The best means of recycling your car's battery is to bring it to a local service station or auto parts dealer that has a means of recycling it or disposing of it safely.

**Corrugated cardboard.** Corrugated cardboard—used primarily in cardboard boxes— like paper, is fully recyclable. Cardboard, in particular, is in short supply, so the demand for used cardboard is high. About half of all corrugated cardboard used in the United States is recovered each year, and the percentage is expected to rise in coming years.

To recycle cardboard, you need only flatten boxes and tie them together for drop-off or pickup.

**Clothing.** Many people overlook the obvious: the clothes on their back. While there are no reliable figures on the amount of reusable worn clothing tossed out each year, common sense would dictate

## *Return to Sender*

Want to eliminate a lot of paper? Consider reducing your mailbox's diet of second- and third-class promotional mail — also known as "junk mail." To reduce junk mail, write to the **Direct Mail/Marketing Association** (6 E. 43rd St., New York, NY 10017) and ask to be eliminated from mailing lists. (Be sure to include the name of all individuals living at your address, including the various ways their names are listed on subscription lists and other mailings.) DM/MA's "mail preference service" will request that your name not be sold by most of the largest mailing list companies. This will elimiate about three-fourths of future promotional mailings going to your address, but won't affect mail you receive from companies that already have your address. Of course, you can still contact select organizations and retailers and ask that your name be added to their list, but that your name not be sold to other organizations. Most organizations will comply with this request.

that any amount is a waste, particularly in light of the millions of Americans who cannot afford to shop for clothes. Humanitarianism aside, these thrown-away clothes represent a tremendous amount of discarded energy and resources.

Almost every community has a local service or religious organization that will pick up used clothing—as well as most other household goods—and distribute them to those in need. Some, but not all, of these organizations repair broken or torn items before distribution.

**Glass.** Like aluminum, glass is 100 percent recyclable. Glass makes up about 10 percent of household garbage and is one of the easiest of materials to recycle. The broken glass (known as "cullet") is added to new molten glass in a furnace, producing new glass. Through the use of cullet, which melts at a lower temperature than the raw materials used to make glass, new glass can be made with considerably less energy and resources than would otherwise be needed. Glass can be reused an infinite number of times. Of course, some bottles needn't be melted down at all. Like milk bottles of years gone by, they are merely washed, sterilized, and refilled.

## *How to Identify Recycled Products*

Finding recycled products at the market takes a bit of investigative work. Very few products boldly announce that their products are made from recycled material, and there have been a couple of instances in which companies' claims have been misleading.

Recycled packaging can be identified in three ways:

- ❑ The recycling symbol on the package (pictured here).
- ❑ A statement such as "This package made from recycled materials."
- ❑ A gray interior in paperboard boxes, such as those used for cereals, detergents, and cake mixes. A white interior usually indicates that the package is made from virgin materials. (However, a box with a white interior may still be made from recycled material. To check, tear a corner of the package; if you see gray, it is made from recycled paper.)

There are three basic types of glass: clear, green, and brown. Not all recycling centers accept all three types, but some do. In any case, you should separate the different-color glass into different containers (or at least separate out the clear glass, if that's all your local recycling center will accept). Broken glass is perfectly acceptable, and while you needn't remove paper labels, you may need to remove aluminum neck rings and caps; your recycling center will provide details. Some types of glass aren't recyclable, however. Most recycling centers, for example, will not accept light bulbs, ceramic glass, dishes, or plate glass, because these items consist of different materials than do bottles and jars.

**Paper.** The United States leads the world in paper consumption—and trails far behind in paper recycling. Americans use some 67 million tons of paper annually, recycling about a fourth of it. This statistic is particularly distressing in light of the fact that the production of a ton of paper from discarded waste paper requires 64 percent less energy, needs 58 percent less water, results in 74 percent less air pollution and 35 percent less water pollution, saves 17 pulp trees, reduces solid waste going into landfills, and creates five times more jobs compared to producing a ton of paper from virgin wood pulp. According to the Institute of Scrap Recycling Industries, more than 200 million trees are saved each year due to current recycling efforts. Paper makes up nearly a third of municipal solid waste by weight and well over half by volume. Each ton of paper not recycled uses three cubic feet of landfill space and incurs as much as $100 in disposal costs.

The lack of paper recycling in America has less to do with people's unwillingness to recycle than with the lack of recycling mills in operation. While the eight mills in the United States capable of recycling newspapers run at full capacity and sell all they produce, the price tag for building a new recycling mill is a whopping $400 million. Considering the fluctuating price of paper—the demand for newsprint, for example, is directly related to the demand for advertising linage, which is linked to the economy as a whole—investors are reluctant to take such a risk.

But the demand for recycled paper exists, and all forecasts indicate that as the demand continues to grow, new manufacturing plants will open.

Not all paper is recyclable. Newsprint is among the most desirable types, but recyclers are picky about even that. Coated papers—such as those used for Sunday magazines and newspaper advertising supplements—can make sheets of rolled paper stick together. Yellow paper, including legal pads, also are undesirable. Brown bags and junk mail also gum up the works. The widespread assumption that brown paper is always recyclable simply isn't true. It must first be separated from other types of paper.

Once you understand these few rules, recycling paper is easy. Newspapers may be kept in a separate bin, then tied (with cotton string) into bundles a foot or two thick for easy handling. Nearly all recycling plants accept bundled newspapers.

Most office papers may be recycled too. Office workers produce an average of a pound of waste paper a day, according to the Office Paper Recycling Service, a New York consulting group. (Ironically, much of the paper is generated by computers and other equipment that was supposed to represent the paperless office of the future.) Unfortunately, the common yellow legal pad is not accepted by most paper recyclers because the color dyes make it difficult to produce the bright white paper most consumers want; other paper colors are equally undesirable.

## *Say What?*

Here are some of the growing number of words and terms used on product labels to indicate their benefits to the environment. But what does each mean? The truth is, none of these means much. All are vague terms that have no agreed-upon or legal definition. You should be wary of these phrases unless you understand exactly what aspects of a product they refer to.

❑**"Nontoxic":** The term is intended to indicate that the product will not have any harmful effect on humans, there is no legal definition of it. And substances that are not poisonous to humans may well harm plant or other animal life. Moreover, toxicity is only one issue. What happens when you dispose of the product? Will its contents or packaging be harmful, even toxic, to the environment? It is difficult to know, based solely on this claim.

❑**"Biodegradable":** *Everything* will biodegrade eventually, even if it takes several thousand years. Again, there is no law mandating what this term is supposed to mean. (See also "The Perils of Plastic.")

❑**"Natural":** This is one of the most abused terms in marketing. There are many "natural" ingredients—lead, for example—that are extremely poisonous.

❑**"Organic":** Although there are some standards for using this term, there is no legal requirement to meet them to make this claim. All living matter and many hazardous chemicals can be said to be "organic." It is best to look for a certifying seal from an established organization. (See "How Safe Is Our Food?")

❑**"Environment friendly," "Environmentally safe," "Won't harm the environment," "Safe for the ozone layer," etc.:** These may be well-meaning phrases, but they are also meaningless. Does the claim have to do with the product's contents, its packaging, or both? Almost any product has some impact on the environment in terms of the resources used to manufacture, package, use, and dispose of it. Be wary of such ideal-sounding but vague terms.

With the demand for recycled office paper being much greater than the demand for newsprint—and, therefore, the price for recycled office paper being considerably higher than for newsprint—many large companies have recognized that recycling office paper is as profitable as it is environmentally sound. According to published reports, the American Telephone & Telegraph Company saved $1 million in disposal costs in 1988 and made a $365,000 profit by recycling its office waste paper. Other companies report similarly impressive figures.

Office recycling programs require central drop-off locations, ideally near frequently traveled locations on each floor, such as the photocopier, the bathroom, or the water cooler. From there, collected paper may be hauled to a central location in the building.

**Oil.** While an increasing number of people are changing their own motor oil, many of them—as well as a number of professional service stations—do not dispose of the used oil properly. And if you think that the few quarts of oil draining from your car's crankcase won't hurt anyone, think again: As little as a quart of oil, when completely dissolved or dispersed in water, can contaminate up to 2 million gallons of drinking water. A single gallon can form an oil slick of nearly eight acres. Used motor oil poured down the drain or into the nearest storm sewer often goes directly to the nearest creek, river, or lake, killing aquatic life and polluting drinking water.

Many communities have begun curbside collection programs to recycle used oil. Used oil is a valuable, renewable resource, although it must be handled carefully. Through refining, about two and a half quarts of new motor oil can be extracted from one gallon of used oil. When recycled, used oil is reprocessed with water, then used for fuel. About a fourth is re-refined and turned into base oil stock. With additives, it is used as lubricating oil and put to other uses. Some of it is used for road oil, dust control, wood preservatives, and "fire log" ingredients. The production of re-refined oil uses just one-fourth the energy of refining from crude oil. According to the U.S. Department of Energy, Americans use about 1.2 billion gallons of oil annually—about 78,000 barrels a day. About 60 percent of it (some 700 million gallons) is used motor oils. The remaining 500 million gallons are industrial oils.

**Plastics.** As stated earlier, plastics recycling is a growing—and controversial—process. (See "The Problem of Packaging.") Researchers have developed processes to shred plastics, primarily from soda bottles, into flakes, which are then washed and separated from residual materials and made into a wide range of materials. But unlike paper, glass, and aluminum, plastics at present can be recycled a limited number of times. Moreover, only a small number of recycling centers or curbside pickup programs currently accept plastics.

**Tin Cans.** Each year, about 30 billion steel cans with a thin tin coating are dumped into America's landfills. The technology to reclaim and recycle these two materials has been around for more than sixty years, and the capacity to recycle far exceeds the availability of recyclable cans. In fact, **Proler International Corporation** (7501 Wallisville Rd., P.O. Box 286, Houston, TX 77001; 713-675-5968), one of the country's largest scrap metal processors, has launched an aggressive advertising and marketing campaign to generate used tin cans for recycling. Proler claims to have the capacity in just one of its plants to recycle 180,000 tons of metal a year, the equivalent of 2.4 billion tin cans.

**Tires.** We throw away 220 million tires a year, millions of which end up in huge unsightly—and fire-prone—stockpiles around the country; currently, well over a billion discarded tires sit in such mounds. But tires can be recycled in a number of ways: almost 40 million tires are retreaded annually; another 10 million are shredded, then used for sheet rubber, asphalt-rubber for roadbeds, and other products; some shredded tires are used as fuel to generate electricity. Some people recycle tires by using them in gardening to surround and protect tomato and other plants.

**Yard Wastes.** Leaves, cut grass, and other yard wastes represent about 20 percent of all waste that ends up in landfills—about 35 billion tons a year. Yet these materials can be of beneficial use, and their recycling can save money. Recycling yard wastes usually takes place in the form of composting them. Composting refers to the biological decomposition of organic waste material (such as plants and food) under controlled conditions. (See the "Garden and Pet Supplies" section for more on composting.)

## Recycling Containers

How do you store your recyclable garbage so that the paper, aluminum, glass, and other materials remain separate and easy to transport? The growth of recycling has created a growing industry in home storage containers to fit the needs of recycling households. Joining the standard issue trash can is a variety of innovative products to make recycling as hassle free as possible.

Most containers designed to handle recycled materials fall into three basic types:

- ❑ **Single containers** usually range from 12 to 22 gallons and are intended for a single type of material.
- ❑ **Stackable multiple container systems** usually consist of three rectangular containers, usually color coded, which range in size from 12 to 14 gallons each. There is typically one container each for newspapers, glass, and aluminum.
- ❑ **Wheeled containers** are the largest and most expensive, ranging from 32 gallons to 105 gallons. These are effectively large wheelbarrows, usually separated into compartments for different materials.

Here is a selected list of home storage container suppliers. Most will supply a catalog or other sales material on request.

Advanced Recycling Systems
P.O. Box 1796
Waterloo, IA 50704
319-291-6007

The Bag Connection
P.O. Box 817
Newberg, OR 97132
503-538-3211
800-622-2448

Bennett Industries
P.O. Box 519
Peotone, IL 60468
312-258-3211

Buckhorn, Inc.
55 W. Techne Center Dr.
Milford, OH 45150
513-831-4402
800-453-4454

Household Recycling Products
P.O. Box 1124
Middleton, MA 01929
508-475-1776

IPL Products Ltd.
348 Park St., Suite 201
North Reading, MA 01864
508-664-5595

LewiSystems
128 Hospital Dr.
Watertown, WI 53094
414-261-4030
800-558-9563

Management Science Applications
123 E. Ninth St., Suite 204
Upland, CA 91786
714-981-0894

Microphor, Inc.
452 E. Hill Rd.
Willits, CA 94590
707-459-5563

Otto Industries
P.O. Box 410251
Charlotte, NC 28241
601-922-0331

Pawnee Products
P.O. Box 751
Goddard, KS 67052
316-794-2213

Philadelphia Can Co.
4000 N. American St.
Philadelphia, PA 19140
215-223-3500

Refuse Removal Systems
P.O. Box 2258
Fair Oaks, CA 95628
916-966-0496
800-231-2212

Rehrig Pacific Co.
4010 E. 26th St.
Los Angeles, CA 90023
213-262-5145

Reuter, Inc.
410 11th Ave. S.
Hopkins, MN 55343
612-935-6921

Ropak Atlantic
2-B Corn Rd.
Dayton, NJ 08810
201-329-3020

Shamrock Industries, Inc.
834 N. 7th St.
Minneapolis, MN 55411
612-332-2100
800-822-2342

Snyder Industries
P.O. Box 4583
Lincoln, NE 68504
402-467-5221

SSI Schaefer
666 Dundee Rd., Suite 1501
Northbrook, IL 60062
312-498-4004

Terc
699 C. Briar Court
Lakehurst, NJ 08733
201-657-4690

Toter, Inc.
P.O. Box 5338
Statesville, NC 28677
704-872-8171

Zarn, Inc.
P.O. Box 1350
Reidsville, NC 27320
919-349-3323

## – Products Made with or Containing CFCs –

Although not every product in each category below contains or is made with chlorofluorocarbons, many of them are. If you think CFCs are hardly being used these days, think again. There are dozens of products still using the chemical, some of which you use every day.

| | Made with | Contains or uses |
|---|---|---|
| **Rigid Foam** | | |
| automotive door panel padding | | ■ |
| boat hull void filler | | ■ |
| buoy, dock protection foam padding | | ■ |
| crack or crevice filler | | ■ |
| disposable dishes | | ■ |
| egg cartons | | ■ |
| extruded polystyrene insulated foam | | ■ |
| fast-food "clam shell" packages | | ■ |
| foam drinking cups | | ■ |
| foam ice chests | | ■ |
| foam packing cushion chips | | ■ |
| freezer insulation | | ■ |
| home insulating sheathing | | ■ |
| refrigerated truck insulation | | ■ |
| rigid pipe insulation | | ■ |
| roof insulation | | ■ |
| storage tank insulation | | ■ |
| Styrofoam | | ■ |
| supermarket meat trays | | ■ |
| **Flexible Foam** | | |
| automobile dashboards | ■ | |
| automobile seat cushions | ■ | |
| bedding foam pillows and mattresses | ■ | |
| bicycle seats | ■ | |
| carpet pads | ■ | |

| | Made with | Contains or uses |
|---|---|---|
| flexible pipe insulation | ■ | |
| furniture cushions | ■ | |
| furniture protective wrapping | ■ | |
| motorcycle seats | ■ | |
| packaging cushions | ■ | |
| postal mailers | ■ | |
| soft toy stuffing | ■ | |
| sports cushioning or pads | ■ | |
| vibration dampeners | ■ | |
| **Refrigeration and Air Conditioning** | | |
| agricultural food chiller | | ■ |
| aircraft air conditioners | | ■ |
| auto air conditioners | | ■ |
| beer dispensers | | ■ |
| bus air conditioning | | ■ |
| cold storage warehouses | | ■ |
| commercial fishing boat refrigeration | | ■ |
| dehumidifiers | | ■ |
| grocery store freezers | | ■ |
| grocery store refrigerators | | ■ |
| home air conditioners, central and window units | | ■ |
| home freezers | | ■ |
| home heat pumps | | ■ |
| home refrigerators | | ■ |
| ice machines | | ■ |
| mall air conditioning | | ■ |
| office building chillers | | ■ |
| refrigerated transport trucks, rail cars | | ■ |
| rooftop and office building air conditioners | | ■ |
| soda fountain dispensers | | ■ |
| soft ice cream or yogurt machines | | ■ |
| truck air conditioning | | ■ |
| vending machines | | ■ |

| | Made with | Contains or uses |
|---|---|---|
| **Electronics, Cleaning Agents** | | |
| air conditioner condenser coils | ■ | |
| artificial hip joints | ■ | |
| automotive electronic components | ■ | |
| ball bearings | | ■ |
| calculators | ■ | |
| cameras | ■ | |
| cardiac pace makers | ■ | |
| clothes dryers | ■ | |
| clothes washers | ■ | |
| computer disc memory storage components | ■ | |
| computers | ■ | |
| contact lenses | ■ | |
| electric and electronic toys | ■ | |
| electronic switches | ■ | |
| garment dry-cleaning fluid | ■ | |
| home heating thermostats | ■ | |
| jewelry | ■ | |
| kitchen dishwashers | ■ | |
| lenses for glasses | ■ | |
| microwave ovens | ■ | |
| photocopiers | ■ | |
| plastic auto body parts | ■ | |
| printed circuit boards | ■ | |
| radios | ■ | |
| rifle scopes | ■ | |
| semiconductors | ■ | |
| silicon wafers | ■ | |
| smoke alarms | ■ | |
| telephones | ■ | |
| television sets | ■ | |
| thermal bulbs for temperature controllers | ■ | |
| thermostats | ■ | |
| typewriters | ■ | |
| VCRs | ■ | |

| | Made with | Contains or uses |
|---|---|---|
| **Miscellaneous** | | |
| aerosol cleaners where exempted* | | ■ |
| aerosol insecticides where exempted* | | ■ |
| aircraft fire extinguisher systems | | ■ |
| blood plasma | ■ | |
| bronchial inhalant medications | | ■ |
| chewing gum remover | | ■ |
| computer disk envelopes | ■ | |
| dusters for cameras | | ■ |
| envelopes for Federal Express | ■ | |
| gum remover | | ■ |
| hemorrhoidal foam | | ■ |
| hospital sterilization | | ■ |
| immersion food freezing | ■ | |
| insecticides | | ■ |
| low-tar tobacco | ■ | |
| marine horns and sirens | | ■ |
| medical apparatus syringes, surgical tubing | ■ | |
| medical sterile packaging | ■ | |
| molded plastic parts | ■ | |
| portable fire extinguishers | | ■ |
| spices | ■ | |
| urethane sole shoes | | ■ |
| weld inspection dye and developer | | ■ |

*The law banning CFC aerosols contained many loopholes, permitting hundreds of products to retain their CFCs. The list includes a wide range of products: high-pressure "dust-off" products used in photography and computer maintenance; boat foghorns; medical supplies; solvents, such as tape recorder head cleaners; and aerosol toy products that create colorful "silly string."

# *Part II*

# S·H·O·P·P·I·N·G F·O·R A B·E·T·T·E·R E·N·V·I·R·O·N·M·E·N·T

# A·U·T·O·M·O·B·I·L·E·S

Cars—we can't live with them or without them. Take them away and some people's entire lives may need to be overhauled. Keep driving them and we may all be suffering from their many health effects on the environment and on our lives.

Is there a happy medium? Is it possible to be a Green Consumer and still enjoy the luxury of independent mobility that our society treasures? Perhaps. A lot of the answer depends on the kinds of cars we drive and on the way we care for them.

Before we explain, let's first take a look at the effects of cars on the environment. You probably already know about air pollution, especially if you live in a big city. It used to be that only a few cities—Los Angeles, Tokyo, and Mexico City, for example—were choking in car exhaust. Now, that list includes more than two dozen cities in the United States alone.

There's no question that automobile exhaust is one of the principal ingredients of environmental problems, including air pollution. In fact, cars and trucks can be implicated directly in three of the biggest crises we face:

❑ **The greenhouse effect:** Cars and light trucks emit one-fifth of all carbon dioxide in the United States, one of the main causes of the greenhouse effect. Burning one gallon of gas produces about twenty pounds of carbon dioxide. For a car driving 1,000 miles a month, that amounts to 120 *tons* of carbon dioxide a year, a not insignificant contribution to the 5.5 billion tons that humans spew into the atmosphere each year. Cars also produce three other gases linked to global warming: nitrous oxide, created when oil is burned; ozone smog, which is a health problem in its own right; and chlorofluorocarbons, which are contained in car and truck air conditioners.

❑ **Air pollution:** Ground-level ozone is created when hydrocarbons and nitrous oxide combine and are exposed to the sun's ultraviolet light. Cars are thought to be responsible for a third of all nitrous oxide and more than a fourth of all hydrocarbons.
❑ **Acid rain:** Again, the culprit is nitrous oxide, one of the major acid-rain precursors.

That's just the beginning of the environmental problems caused by cars and trucks. Consider the trash: some 200 million tires and 64 million automobile batteries are thrown away each year, each contributing its share of pollutants. (See "The Recycling Solution" for more information about tires and batteries.) And then there are the cars themselves. The 10 million or so cars sold in the U.S. each year consume nearly 9 billion tons of steel, plus vast quantities of other raw materials. Close to 9 million cars are junked each year, only a fraction of which are recycled as scrap metal. The rest fill our landfills or litter our landscape.

None of this includes the impact of oil exploration and drilling on the environment, and the oil spills that all too frequently result, wreaking havoc on every plant, animal, and organism for hundreds of miles around. Cars and light trucks guzzle more than a third of all oil used in the United States.

Why are cars so polluting? Haven't we enacted clean-air laws that have mandated automobile fuel-efficiency standards? Aren't cars getting more environmentally sound?

True, federal law has dictated certain fuel-efficiency requirements for two decades, since the enactment of the Clean Air Act of 1970. But a funny thing happened on the road to fuel efficiency and cleaner air: The number of cars on the road has surged more than 50 percent since 1970, to about 135 million. One problem is that from almost the day the clean-air law was enacted, American automobile manufacturers have fought the fuel-efficiency standards, often succeeding in weakening them. (Detroit's argument is to wait things out: The environment will clean itself up as more consumers replace their older, gas-guzzling cars with newer, more efficient ones. Clearly, automobile manufacturers have failed to learn the lesson of the past two decades: More cars create more pollution, no matter how efficient they may be.) As a result, even with more fuel-efficient cars, the nation's air is far dirtier than when the landmark law was passed.

So dirty, in fact, that states and other local jurisdictions are no longer waiting for Uncle Sam to come to terms with the nation's auto makers. Officials in southern California, for example, created a tough plan to sharply curtail automobile use around Los Angeles. In the Northeast, the six New England states plus New York and New Jersey have banded together to require auto makers to meet emissions standards tougher than those allowed by federal law. In 1989, the Vermont legislature overwhelmingly passed a bill outlawing automobile air conditioners by 1993 unless a new coolant is substituted for ozone-damaging chlorofluorocarbons.

## A Greener Car?

It's not that car makers aren't capable of building a more fuel-efficient, less-polluting car. They need only follow the technology pioneered by the Japanese, such as multi-valve engines (which have four valves per cylinder instead of two) and continuously variable transmissions (which lack gears, instead of adjusting constantly to a car's changing speed). It is these and other features that have long allowed the Japanese to hold the number-one position in fuel economy for cars sold in the United States. (The Chevrolet Geo, the most fuel-efficient of all 1990 models—imports or domestics—which is rated at 53 miles per gallon in the city and 58 miles per gallon on the highway, is imported from Japan; see gas mileage ratings on page 75.) According to a report by the federal government's Office of Technology Assessment, such technology would help American auto makers achieve a 33-miles-per-gallon average—up considerably from today's 27.5 mpg—"without performance loss or the need to move to smaller vehicles." Additional costs of these innovations would be more than covered by increased fuel efficiency.

So despite the many controversies surrounding finding new sources of oil, it can be said that America's largest "oil reserves" have no environmental impact: They can be achieved by making fuel-efficiency improvements in American-built cars.

Not that American car makers aren't testing new technologies of their own. Since the mid-1970s, Detroit auto makers—spurred on by strong pressure from the federal government—have experimented with a variety of vehicles that run on little or no gasoline, opting for any of several less-polluting alternative fuels.

Here are some of the most promising prospects for alternative fuels during the 1990s:

❑ **Gas-methanol engines:** Most of the cars being tested now use a combination of 15 percent gasoline and 85 percent methanol, an alcohol derived from natural gas, wood, and other sources. Auto racers use methanol because it produces better engine performance.

❑ **Ethanol:** Produced from corn, barley, rye, wheat, sugar cane, sugar beet, and other crops, this fuel already powers 2 million vehicles in Brazil—most of them manufactured by General Motors and Ford. According to Columbia University professor Harry P. Gregor, "If ethanol use were mandated for all city-owned or regulated vehicles such as buses, taxis, police cars, and garbage trucks in the nine U.S. areas of highest air pollution, air quality would rapidly and markedly improve." The cost of changing to ethanol was tremendous, however, and initially led to worsening the air pollution problem in Brazil, due to misadjustments in engine conversions that created high evaporations. A new problem, especially in northeastern Brazil, arose when sugar prices skyrocketed and the price of sugar beets rose as well, producing a shortage of ethanol.

❑ **Flexible-fuel engines:** Many experts predict that when alternative-fuel cars do come out—probably toward the end of the decade—they will be able to use either methanol or gasoline, permitting motorists to opt for the most available and affordable fuel at any given moment. In 1986, both Ford and General Motors introduced special models that can run on gas, methanol, and several other fuels—singly or in combination. (The cars use an optical scanner, which analyzes what's in the fuel line and regulates the fuel injectors accordingly.) But neither Ford's "flexible fuel" Crown Victoria nor G.M.'s "variable fuel" Chevy Corsica is yet being mass produced.

❑ **Natural gas:** This represents one of the best prospects for fueling trucks. Currently, about 30,000 vehicles powered by natural gas are in use around the country, according to the Natural Gas Vehicle Coalition. United Parcel Service, the world's largest package delivery company, is undergoing a two-year experiment that eventually could result in a large portion of the company's 104,000 vehicles switching to natural gas.

Each of these has distinct advantages, along with some disadvantages. For example, because they are more corrosive than gasoline, methanol and ethanol would require that cars be equipped with a stronger fuel pump and sturdier coatings on some engine parts. Methanol exhaust contains higher levels of the poisonous gas formaldehyde, although car makers are experimenting with devices to remove formaldehyde from methanol exhaust. But methanol produces few hydrocarbons or nitrous oxides—the two principal contributors to urban smog, as well as the greenhouse effect, in the case of nitrous oxides.

None of these—or any other—alternatives will work, however, without an abundance of these fuels as close as the corner service station. Indeed, the lack of widespread availability of these fuels at service stations may be the biggest obstacle of all for alternative fuels. Oil company resistance to having to produce and market alternative fuels along with gasoline and diesel has slowed progress considerably.

The oil companies would rather Americans stick with gasoline—albeit a cleaner, more efficient gasoline. Indeed, much of the antipollution efforts of some oil companies—most notably Chevron Corporation and Atlantic Richfield Company's ARCO—have focused on reformulated gasoline. ARCO, for example, has hyped its new "EC-1 Regular" (the "EC" stands for "emission control"), which it introduced in southern California in 1989. Intended for the heaviest-polluting vehicles—pre-1975 cars and pre-1980 trucks that run on leaded gasoline and do not have emissions-reducing catalytic converters—the fuel was claimed by ARCO to cut automobile pollution substantially, without affecting performance or costing consumers extra; actual evidence is inconclusive. According to ARCO, "If all leaded gasoline users in southern California switched to a gasoline like ARCO EC-1, about 350 tons of pollutants would be removed from the air each day. The impact would be equivalent to permanently removing 20 percent—or more than 300,000—of these high-polluting, pre-catalyst-equipped vehicles from southern California highways."

Ultimately, the oil companies must give in to the realities of nonpetroleum-based fuels. There is no foreseeable technology that will reduce automobile pollution to acceptable levels if gasoline is used. While improvements will continue to be made in both automobiles and their fuels, most experts agree that the time is not far away when a sizable portion of vehicles will operate on alternative fuels.

That day will arrive even sooner if the federal government is willing to pass tough new fuel-mileage rules. So far, the debate between car makers and law makers is over a relatively small range—whether the average mileage for all cars produced by a given manufacturer be 26.5 or 27.5 or 28.5 miles per gallon. But those arguments obscure the real point: Most experts believe that the technology exists to raise the average to 35 or even 45 mpg. Several car makers have built prototype cars with impressive fuel efficiency—Volvo's LCP 2000 gets 65 mpg, Volkswagen's E80 gets 74 mpg in the city and 99 mpg on the highway, and Toyota's AXV gets a whopping 89 mpg in the city and 110 mpg on the highway. But such cars inexplicably never seem to reach the public.

## How to Be a Greener Driver

For Green Consumers, the real bottom line is this: There are no green cars or green fuels. Driving pollutes, no matter what you drive, how you drive it, what the fuel efficiency is, or what kind of fuel you use.

That doesn't mean that you can't be a *greener* driver, reducing your own car's contribution to environmental problems. Here are nine relatively simple ways you can help:

❑ **Drive fewer miles, or more miles per gallon.** This doesn't necessarily mean that you must drive less or purchase a more fuel-efficient car, although both of these steps would help even more. If you think of "miles per gallon" as the miles traveled on a gallon of fuel multiplied by the number of people carried, you can increase your mpg simply by bringing more people along for the ride. Combine errands; share trips with a neighbor.

For example, one person driving to work in a 20-mpg car equals 20 miles per gallon, while three people driving to work in that same car get 60 mpg. Those same three people driving in a 30-mpg car would achieve 90 miles per gallon. Thirty people riding to work on a fuel-efficient diesel bus might achieve over 150 miles per gallon. Form car pools with neighbors and co-workers.

## *Fuel Economy and Carbon Dioxide*

The better your car's gas mileage, the less carbon dioxide your car will emit into the environment. As stated earlier, cars and light trucks emit one-fifth of all carbon dioxide in the United States, one of the main causes of the greenhouse effect.

Here are estimates of the amount of carbon dioxide emissions over a single car's lifetime, courtesy of the Energy Conservation Coalition, a project of Environmental Action:

### *Keeping the Pressure Up*

Who says the little things don't count? Sometimes, as we view the enormity of the environmental problems of our world, we forget about how some of the little things can make a big difference.

Consider automobile tire pressure, for example. If you are a typical American driver, you probably are driving around with underinflated tires. That may not seem like a major environmental blunder, but in fact it is making a significant contribution to global warming. The basic assumption has to do with the fact that each gallon of gas your car uses contributes about 19 pounds of carbon dioxide ($CO_2$) into the atmosphere. (See "Seven Environmental Problems You Can Do Something About" for more on the link between $CO_2$ and global warming.)

According to the Environmental Protection Agency,at least 65 million cars on American roads have underinflated tires. That low air pressure increases a car's rolling resistance, decreasing gas mileage by up to 5 percent. If you assume that these 65 million cars get an average of 18 miles per gallon (that's the EPA's estimate), that each drives 10,000 miles per year, and that 5 percent of the gas is wasted due to underinflated tires, that comes to *17 million tons* of $CO_2$ released into the air simply because of low tire pressure.

So, keep those tires inflated according to the manufacturer's recommendations. It seems such an easy way to make a difference.

❑ **Keep your car properly maintained.** A poorly tuned car can use as much as 10 percent more gasoline than a well-tuned one. The cost of the tune-up will be partially offset by the increased gas mileage your car will get. The Environmental Protection Agency recommends that cars built before 1981 get regular tune-ups. However, most post-1981 models are built so that an annual tune-up is not necessary and may be an unneeded expense. EPA does suggest annual *inspections* to make sure everything is running properly.

Among the other maintenance tasks that can save gas:

—Keep the engine filters clean. Clogged filters waste fuel.

—Check tire pressures regularly. Underinflated tires increase gas use. (See box above.)

—Keep the brakes adjusted. Dragging brakes can reduce fuel efficiency as well as wear down the brakes themselves.

—Change the oil regularly. Keeping the engine oil clean maintains engine life, ensuring its maximum efficiency. Use a high-quality multigrade oil (such as 10W-30 or 10W-40) to reduce engine friction and increase fuel efficiency.

— Check your wheel alignment. A car with the front wheels out of alignment will waste fuel.

❑ **Consider radial tires.** They can result in a 3 to 5 percent improvement in gas mileage in the city, 7 percent on the highway, and 10 percent at 55 mph after the tires have been warmed up for 20 minutes. Radials last longer, too. (Never mix radials with conventional tires without consulting your car owner's manual.)

❑ **Remove unnecessary weight from the car.** The lighter the car, the less gas it uses. An extra 100 pounds decreases fuel economy about 1 percent for the average car, slightly more for small cars.

❑ **Don't idle.** Don't idle the engine; your car gets 0 miles per gallon when it is idling. Once the car is warmed up (after about five minutes of driving, slightly more in the dead of winter), it's better to stop and restart it.

❑ **Obey speed limits.** Your car gets 20 percent better mileage at 55 mph than at 70 mph.

❑ **Minimize your use of the air conditioner.** Besides being a major source of ozone-destroying chlorofluorocarbons (more on that in a moment), air conditioning reduces fuel economy by up to 2.5 miles per gallon. When purchasing a car, consider a light-colored car with tinted glass, two factors that will keep the car cooler in summer, requiring less air conditioning.

❑ **Change your driving behavior.** Drive smoothly; accelerate and slow down gradually. Don't take a lot of short trips. By planning your trips carefully, you'll save fuel. The less city driving you do, the better. The average speed in congested downtown areas can cut your fuel efficiency by more than half. And there is no need to rev your engine just before shutting it off. This practice is not necessary for today's cars; it only wastes fuel.

❑ **Don't drive when there's a reasonable alternative.** Of course, not driving at all is the best way to save gas and reduce air pollution. Walking, biking, and taking public transportation are sure-fire ways to improve the environment.

## Disposing of Hazardous Automobile Substances

Your car is a repository of a variety of fluids that can harm the environment: gasoline, gas additives, motor oil, transmission fluid, brake fluid, antifreeze, coolant, and battery acid. If you do some of your car's maintenance work yourself (a lot of people do—25 percent of car owners change their own oil), the way you dispose of these fluids can have a severe environmental impact. Here is the proper way to dispose of common automobile products:

❑ **Motor oil, brake fluid, and transmission fluid:** Don't dump these down the drain or even the sewer. Instead, drain these used fluids into a sturdy, sealable container and take it to your neighborhood garage. Protect used oil from water and debris; contaminated oil won't be accepted for recycling. Most garages have environmentally acceptable methods of disposing of or recycling these fluids. There may also be a recycling pickup point, oil collection center, or hazardous-waste collection site in your area that accepts these substances. (See "The Recycling Solution" for more on recycling motor oil.)

❑ **Antifreeze/coolant:** Antifreeze contains glycol, which is poisonous to people, fish, and wildlife. Many family pets have died after drinking sweet-tasting puddles of antifreeze they find on driveways or in the street. Pour used antifreeze into a sturdy container and bring it to a local service station; ask the attendant to add the liquid to the station's antifreeze storage drum. You may also dilute the antifreeze with water and pour the mixture into a gravel pit or any area with good drainage. If the coolant has been used in an older car, it may contain dangerous levels of lead. Take such coolant to a service station or to a hazardous-waste dump.

❑ **Gasoline:** If the gasoline can be used in your car, use it. If it is contaminated, bring the gas to a hazardous-waste collection site as soon as possible. By all means, do not mix gasoline with waste oil, or anything else, unless instructed to do so by a product label. Keep in mind that storing gasoline at home is *extremely* dangerous. If you must store it, use only containers designed for that purpose.

❑ **Batteries:** Car batteries contain lead and sulfuric acid, both of which are extremely dangerous. These batteries are considered to

## *Automotive Hazardous-Waste Disposal*

| Product | Hazardous Ingredients | Alternatives | Hazardous Properties | Disposal Method |
|---|---|---|---|---|
| Antifreeze | ethylene glycol | unknown | toxic | 1, 2 |
| Batteries | sulfuric acid, lead | pedal power | corrosive, toxic | 1 |
| Brake Fluids | glycol ethers | | | |
| | heavy metals | unknown | flammable, toxic | 1 |
| Transmission Fluid | hydrocardbons | | | |
| | mineral oils | unknown | flammable, toxic | 1 |
| Used Oils | hydrocarbons | | | |
| | heavy metals | unknown | flammable, toxic | 1 |

**Disposal Notes:**

1. Recyclable. Take to a service station, reclamation center, or household hazardous-waste collection center. Partially used useful products such as paints may be exchanged.
2. Waste can be disposed of at some wastewater treatment plants where bacteria can detoxify the chemical. (Call your local treatment plant.) Do not pour on the ground.

Courtesy Environmental Hazards Management Institute, 10 Newmarket Rd., P.O Box 932, Durham, NH 03824; 603-868-1496

be one of the most hazardous toxics to the environment if not disposed of properly. Car batteries also can be recycled. (See "The Recycling Solution" for details.)

## CARS AND CFCs

The use of chlorofluorocarbons (CFCs) is one of the biggest environmental problems facing the earth (see "Seven Environmental Problems You Can Do Something About"). While we have succeeded in banning CFCs from most aerosol cans and other applications, cars remain one of the biggest sources of CFC emissions. (Although aerosol cans were banned in the mid-1970s, there are several loopholes in the law, and some CFC-based aerosol products are still on store shelves. Many of these are automobile maintenance products.)

CFCs are the principal refrigerant used in the 90 million cars and light trucks in the United States with air conditioning. Most commonly used is CFC-12 (also known as R-12), which happens to be the type of CFC most destructive to the ozone layer. The Environmental Protection Agency has estimated that CFCs in cars alone account for as much as 25 percent of CFC use in the United States; lost coolant from cars account for an estimated 20 percent of ozone destruction.

The CFCs in a car's air conditioner are contained in a closed system—that is, the refrigerant recycles continuously throughout the air-conditioning system and does not get consumed like gas or oil. If all the CFCs remained safely contained in the system, it would have little impact on the environment.

Of course, the world isn't that simple. The refrigeration systems within car air conditioners leak, releasing CFCs into the atmosphere. Sometimes the entire air-conditioning system breaks down and at the service station, CFCs are removed—usually by releasing them into the air—before the system is repaired. Even if the air conditioner never breaks down, in time the car will, and when it is finally junked, the air conditioner's CFCs will eventually find their way out of the system.

Contributing to the problem has been the approach to air-conditioning maintenance prevalent among car owners and service technicians. One common repair—often made annually—is to have the car's coolant recharged. But recharging indicates a leak in the system, a leak that will not go away regardless of how many times you recharge the coolant. So this "repair" is really a sham—and a horrible affront to the environment.

The auto repair industry has only recently recognized the huge contribution it has been making to ozone depletion. Indeed, the Mobile Air Conditioning Society, a trade association, has requested that its members post the following notice in their shops' waiting rooms:

> Please don't ask us to recharge your leaking A/C system without making necessary repairs. Any refrigerant remaining in your system, along with any refrigerant added, will eventually escape. Such emissions are destructive to our environment. We would rather lose your business than knowingly pollute the atmosphere for the sake of convenience.—*The Management*

In a letter to its members, the society instructed technicians to "explain the situation to your customer, and if he still insists on 'just a little shot of refrigerant,' courteously send him on his way—without it."

But recharging air conditioners is only part of the problem. As stated earlier, when air conditioners are serviced, they are first "bled"—that is, the CFCs are discharged into the air. This practice is changing, slowly. A growing number of service stations are using recovery and recycling equipment to capture and reprocess CFCs. Simply put, these devices pump the refrigerant out of the compressor, clean and purify it for reuse, then store or transfer it while an air conditioner is being repaired. Some states have passed laws that will mandate the use of such equipment in coming years; the federal government, too, is considering such a law.

Several companies manufacture these CFC recovery devices—which have come to be known as "vampires," probably because they suck vital fluids out of air-conditioning systems—many of which cost under $1,000. The good news for service stations is that these devices eventually will pay for themselves in refrigerant savings; by reusing existing coolant, technicians will need to add less fresh coolant to the system. Still, service stations have been slow in purchasing such technology, largely because there has been little consumer demand for it.

Eventually, other, less harmful coolants will be created for air conditioners. Until that happens, you must take special care when it comes to the servicing of your car's system. Patronize only those service stations that use refrigerant reclaiming equipment that is tested by Underwriters Laboratories. To encourage use of these devices, you might even decide to boycott gas stations and repair shops that don't use this equipment, even if you're not bringing your car in for air-conditioning problems. Your voice—and especially your dollars—can be a persuasive force in getting service stations to adopt environmentally responsible practices.

## The Most Fuel-Efficient Cars

Below is a summary of the top and bottom of the fuel-efficiency ratings for 1990 model cars, as published by the U.S. Department of Energy. The figures below list only the top-rated models—those with annual fuel costs under $500 for passenger cars and $650 for light trucks. The annual fuel costs are based on driving 10,000 miles per year in a mix of city and highway driving. The figures are only estimates to be used for comparison figures. As you probably have heard many times, "Your actual mileage may vary." This, of course, has to do with the many factors affecting fuel economy. (See "How to Be a Greener Driver.")

Still, these ratings are relatively accurate to each other. They probably are somewhat overstated compared to "real-life" driving for all of the models, but you can still compare cars equally.

As you'll see, each car listed includes its engine size (in cubic liters), the number of cylinders (the most fuel-efficient engines have either three or four cylinders), and type of transmission and number of gears ("M5," for example, means a five-speed manual transmission).

The complete fuel-efficiency ratings of current-year models is released each fall by the federal government. To obtain the most recent ratings, write to **Environmental Protection Agency**, Public Information Center, 401 M St. SW, Washington, DC 20460; 202-382-2080. The latest mileage ratings, along with a great deal of other useful information about new car models, is available in the authoritative annual *The Car Book*, by Jack Gillis. To obtain the latest edition, send $10.50 (including $1.55 shipping) to the **Center for Auto Safety** (2001 S St. NW, Washington, DC 20009; 202-328-7700), or it can be found for $8.95 in bookstores, published by Harper & Row.

## *Passenger Cars*

| **Car** (Engine size/cyl./trans.) | **City** | **Highway** | **Fuel Cost** |
|---|---|---|---|
| Geo Metro X (1.0L/3/M5) | 53 | 58 | $ 287 |
| Honda Civic CRX HF (1.5L/4/M5) | 49 | 52 | $ 315 |
| Chevrolet Sprint (1.0L/3/M5) | 46 | 50 | $ 335 |
| Geo Metro (1.0L/3/M5) | 46 | 50 | $ 335 |
| Geo Metro LSI (1.0L/3/M5) | 46 | 50 | $ 335 |
| Pontiac Firefly (1.0L/3/M5) | 46 | 50 | $ 335 |
| Suzuki Swift (1.0L/3/M5) | 46 | 50 | $ 335 |
| Volkswagen Jetta (1.6L/4/M5) | 37 | 43 | $ 338 |
| Honda Civic CRX HF (1.5L/4/M5) | 43 | 49 | $ 350 |
| Suzuki Swift (1.3L/4/M5) | 40 | 44 | $ 375 |
| Chevrolet Turbo Sprint (1.0L/3/M5) | 37 | 42 | $ 403 |
| Daihatsu Charade (1.0L/3/M5) | 38 | 42 | $ 403 |
| Pontiac Turbo Firefly (1.0L/3/M5) | 37 | 42 | $ 403 |
| Chevrolet Sprint (1.0L/3/A3) | 38 | 39 | $ 414 |
| Ford Festiva (1.3L/4/M5) | 35 | 41 | $ 414 |
| Geo Metro (1.0L/3/A3) | 37 | 40 | $ 414 |
| Geo Metro LSI (1.0L/3/A3) | 37 | 39 | $ 414 |
| Pontiac Firefly (1.0L/3/A3) | 38 | 39 | $ 414 |
| Suzuki Swift (1.0L/3/A3) | 38 | 39 | $ 414 |
| Daihatsu Charade (1.3L/4/M5) | 35 | 39 | $ 425 |
| Ford Escort (1.9L/4/M4) | 32 | 42 | $ 438 |
| Honda Civic (1.5L/4/M4) | 33 | 37 | $ 450 |
| Pontiac LeMans (1.6L/4/M5) | 31 | 40 | $ 450 |
| Subaru Justy (1.2L/3/M5) | 33 | 37 | $ 450 |
| Isuzu Stylus (1.6L/4/M5) | 32 | 37 | $ 463 |
| Subaru Justy (1.2L/3/AV) | 34 | 35 | $ 463 |
| Dodge Colt (1.5L/4/M4) | 31 | 36 | $ 477 |
| Geo Storm (1.6L/4/M5) | 31 | 36 | $ 477 |
| Honda Civic (1.5L/4/M5) | 31 | 34 | $ 477 |
| Honda Civic CRX (1.5L/4/M5) | 32 | 35 | $ 477 |
| Honda Civic Wagon (1.5L/4/M5) | 31 | 34 | $ 477 |
| Mitsubishi Mirage (1.5L/4/M4) | 31 | 36 | $ 477 |
| Plymouth Colt (1.5L/4/M4) | 31 | 36 | $ 477 |
| Toyota Tercel (1.5L/4/M4) | 31 | 36 | $ 477 |
| Daihatsu Charade (1.3L/4/A3) | 31 | 34 | $ 491 |
| Ford Festiva (1.3L/4/A3) | 31 | 33 | $ 491 |
| Hyundai Excel (1.5L/4/M5) | 29 | 36 | $ 491 |
| Hyundai Precis (1.5L/4/M5) | 29 | 36 | $ 491 |
| Mazda 323/323 Protege (1.6L/4/M5) | 29 | 37 | $ 491 |

| **Car** (Engine size/cyl./trans.) | **City** | **Highway** | **Fuel Cost** |
|---|---|---|---|
| Nissan Sentra (1.6L/4/M4) | 29 | 36 | $ 491 |
| Pontiac LeMans (1.6L/4/M4) | 29 | 38 | $ 491 |
| Subaru Justy 4WD (1.2L/3 AV) | 32 | 32 | $ 491 |
| Suzuki Swift (1.3L/4/A3) | 31 | 34 | $ 491 |
| Suzuki Swift GT (1.3L/4/M5) | 29 | 36 | $ 491 |
| Toyota Tercel (1.5L/4/M5) | 30 | 36 | $ 491 |

## *Light Trucks*

| **Car** (Engine size/cyl./trans.) | **City** | **Highway** | **Fuel Cost** |
|---|---|---|---|
| Suzuki Samurai Hardtop (1.3L/4/M5) | 28 | 29 | $ 562 |
| Suzuki Samurai Soft-Top 4WD (1.3L/4/M5) | 28 | 29 | $ 562 |
| Geo Tracker 4WD (1.6L/4/M5) | 26 | 28 | $ 583 |
| Geo Tracker Conv. 4WD (1.6L/4/M5) | 26 | 28 | $ 583 |
| Suzuki Sidekick 2WD (1.6L/4/M5) | 27 | 28 | $ 583 |
| Suzuki Sidekick Hardtop 4WD (1.6L/4/M5) | 26 | 28 | $ 583 |
| Suzuki Sidekick Soft-Top 4WD (1.6L/4/M5) | 26 | 28 | $ 583 |
| Ford Ranger Pickup 2WD (2.3L/4/M5) | 24 | 29 | $ 606 |
| Subaru Loyale 4WD (1.8L/4/M5) | 24 | 29 | $ 606 |
| Subaru Loyale Wagon 4WD (1.8L/4/M5) | 24 | 29 | $ 606 |
| Chevrolet S10 Pickup 2WD (2.5L/4/M5) | 23 | 27 | $ 630 |
| GMC S15 Pickup 2WD (2.5L/4/M5) | 23 | 27 | $ 630 |
| Nissan Hardbody 2WD (2.4L/4/M5) | 23 | 27 | $ 630 |
| Plymouth Colt Wagon 4WD (1.8L/4/M5) | 23 | 28 | $ 630 |
| Suzuki Sidekick Hardtop 4WD (1.6L/4/L3) | 24 | 25 | $ 630 |
| Suzuki Sidekick Soft-Top 4WD (1.6L/4/L3) | 24 | 25 | $ 630 |
| Toyota Truck 2WD (2.4L/4/M4) | 23 | 28 | $ 630 |

# F·O·O·D AND G·R·O·C·E·R·I·E·S

The local supermarket has increasingly become an exciting but confusing place. The excitement comes from the never-ending stream of new products, each with some new ingredient, package, or marketing claim that sets it off from the competition. It's this very set of circumstances that makes the supermarket confusing, too.

Product manufacturers—whether of food or nonfood items—have become increasingly daring in their sales pitches. They'll boldly claim a product has no cholesterol when it is well known by nutritionists that such a product, while indeed containing no cholesterol, may be very high in saturated fat, which has been strongly linked to heart disease. So the claim, though technically correct, is quite misleading to someone concerned about heart disease. "Natural" and "light" are among the other trendy words that have come to be overused—and abused—by product manufacturers hoping to catch a wave of consumer interest.

As green products are of increasing interest to consumers, some manufacturers have become quick to label a product "biodegradable," "environment friendly," or by some other term, when that isn't necessarily the case. (See "Say What?" page 50.) And there are some perfectly green products out there about which little is said in regard to their environmental benefits.

## Green Packaging, Green Products, Green Grocers

What are the big issues at the supermarket, environmentally speaking? For Green Consumers, there are two principal concerns:

❑ **Packaging:** We've already mentioned some of the key issues related to packaging and overpackaging. Nowhere is this issue greater than at the supermarket, where we buy more "packaged goods" (as they are known in the industry) than anywhere else. This is the source of the most plastic packaging, nonreturnable bottles, and nonessential products. This is where we purchase the vast majority of products destined to become household waste. All of which we pay for dearly—when we purchase these products and when we throw them away. Packaging issues relate both to food and nonfood products.

❑ **Products:** There are a number of products, particularly nonfood products, that contain ingredients harmful to the environment. A few of these—some laundry detergents, for example—contaminate the air and water as they are used. Other products do their damage when they are thrown away. In most cases, there are environmentally preferable alternatives.

Beyond these issues is the issue of the supermarkets themselves: How much do their policies and practices contribute to environmental problems? Some stores, recognizing that they themselves produce a great deal of trash (cardboard boxes, spoiled food, etc.) have taken a hard look at the way they conduct business. These stores' managers recognize that being environmentally responsible is not only good for the planet, it is also good for business. Green Consumers are going out of their way to patronize such stores—and to avoid those stores that are not taking positive action on green issues.

A few supermarkets are taking bold steps. Some examples:

❑ **Giant Food Stores**, for example, a 147-store chain located around the Washington, D.C., area, recycles all of its cardboard. It also has more than 50 newspaper drop-off centers and eight aluminum recycling centers. Giant is in the process of setting up plastic and bottle recycling centers as well.

❑ **Bread and Circus,** a Boston supermarket chain, has successfully sold only food grown without pesticides and drugs.

❑ **Loblaw's,** a Canadian chain, has introduced a "President's Choice" line of "green" products. Included in the line are unbleached coffee filters, products made of recycled paper, phosphate-free laundry detergent, and baking soda marketed as a low-impact, nontoxic household cleaner. (Truth be told, not all of Loblaw's green products are *truly* green. For example, one of the products is a set of disposable paper picnic plates. While paper products are better than plastic products, Green Consumers should question the need for these disposable, nonreusable plates in the first place!)

A positive development took place in 1989, when five American and Canadian supermarket chains pledged to stop selling by 1995 fresh fruits and vegetables treated with cancer-causing pesticides, the first step of a campaign to reduce reliance on agricultural chemicals. The stores include **Provigo,** which operates more than 1,000 stores in Canada; **Petrini's Supermarkets,** a subsidiary of Provigo, which has 23 stores in the San Francisco area; **Raley's Supermarkets,** with 57 stores in the Sacramento, California, area; **ABCO Supermarkets,** with 87 stores in Arizona; and **Bread and Circus,** with five stores in Boston. Ironically, the Environmental Protection Agency, which has been mired for years in its own efforts to rid the market of unsafe pesticides, criticized the announcement for effectively shifting decision-making power to retailers, who the Environmental Protection Agency believes lack the appropriate scientific information. Your support of these stores—and your comments to your own grocer to join the campaign—will help send a message encouraging other supermarkets to do their part to rid the marketplace of environmentally undesirable substances.

Still, these 172 stores represent a tiny fraction of the nation's approximately 148,000 grocery stores—not to mention drugstores, convenience stores, general stores, and all the other retailers that sell food and groceries. (Restaurants are another matter altogether. In this chapter, we'll deal mainly with fast-food outlets. Determining the greenness of other restaurants is somewhat more difficult.

*[continued on page 82]*

## *The Green Hall of Fame*

Here are some grocery products we believe are worthy of special attention because of the efforts their companies have made to apply Green Consumer principles to contents and packaging — and, in some cases, to the operation of the companies themselves. While not all of these products are perfect solutions to environmental problems, they represent a cross section of efforts by large and small companies to address environmental concerns. You are encouraged to buy these products whenever possible to support their manufacturers' pioneering efforts at Green Consumerism. In future editions of this book, we hope to "induct" other trendsetting products into our Green Hall of Fame.

**C.A.R.E. products** (Ashdun Industries, Inc., 1605 John St., Fort Lee, NJ 07024; 201-944-2650) manufactures "environmentally friendly" household products, including paper towels, toilet paper, facial tissue, napkins, coffee filters, cotton swabs, and cotton balls. The C.A.R.E. (for "Consumer Action to Restore the Environment") products are made entirely from recycled fibers using a chlorine-free bleaching process, making them preferable to other brands.

**Downey Fabric Softener Refills** (Procter & Gamble, One P&G Plaza, Cincinnati, OH 45202; 513-983-1100) is among the first of what is expected to be a major trend in packaging: refillables. After using up a 64-ounce plastic jug of Downey, you then buy the refill, packaged in a 21½-ounce milk carton-like cardboard container. Adding two cartons of warm water makes 64 ounces of full-strength fabric softener. The plastic container is infinitely reusable.

**Fountain Fresh Beverage Company** (2030 N. Redwood Rd., Suite 70, Salt Lake City, UT 84116; 801-538-0060) offers refillable soft drink products in the Pacific Northwest and Midwest. Two-liter bottles of Fountain Fresh soda (which comes in more than two dozen flavors) may be returned to selected grocery stores for refilling. A machine automatically washes the bottle and refills it. The initial bottle is 99 cents; refills are 69 cents each.

**Melitta Natural coffee filters** (Melitta U.S.A., 1401 Berlin Rd., Cherry Hill, NJ 08003; 609-428-7202, 800-451-1694) are unbleached, which helps to avoid contamination of dioxins found in many bleached paper products. Melitta has demonstrated that consumers do not demand perfectly white paper products; they can be satisfied with less-than-pristine paper goods, if these are safer for health and the environment.

**Pillsbury Microwave cake mixes** (The Pillsbury Company, 200 S. 6th St., Minneapolis, MN 55402; 612-330-4966) are the first reusable bakery product. Whereas other convenience bakery products require you to throw away the used cake pan and purchase another one, this product includes both cake mix and a reusable pan. After the first purchase, you need buy only refills. While the ideal solution would be to use your own glass baking dish, Pillsbury has come up with a reasonable alternative.

**Scotch Corporation concentrates** (P.O. Box 4466, 617 E. 10th St., Dallas, TX 75208; 214-943-4605) sells all-purpose cleaner and glass-cleaner concentrates in small plastic pouches. When mixed with water in a 22-ounce spray bottle (the size commonly used for leading household spray products) they become convenient, familiar cleaners; you can reuse the plastic bottle indefinitely, cutting down on plastic trash. Because most such cleaners are 90 percent water, this not only saves packaging but reduces the energy used in transportation.

**Wm. T. Thompson Company vitamins** (23529 S. Figueroa St., P.O. Box 6201, Carson, CA 90749; 213-830-5550) demonstrate that natural vitamins can also be good to nature. The company abandoned plastic containers for glass and replaced polyurethane packaging materials for recycled ones. Moreover, the company has involved itself in a wide range of environmental issues, including sponsoring environmental television programs and funding an "earth-friendly" agricultural program in northern California.

Your best bet is to consider disposables: Is food served in foam dishes containing CFCs? Does the restaurant seem wasteful in other ways?)

Someday, choosing green products at the supermarket will be much easier than it is today. American stores will provide shoppers with some of the same environmental amenities offered by their counterparts in Europe: green products, green shelf labeling, and green product labeling. That day will come sooner if Green Consumers raise their voices and demand this information.

For now, let's examine some of the things you should know about food—and what you can do today to make your trip to the market a little greener.

## HOW SAFE IS OUR FOOD?

We used to take food safety for granted. Only when there was some kind of scare—often in the form of a food-tainting episode created by one or more "terrorists"—did we worry about what was in our food.

Perhaps we should have worried about it a bit more often. Despite federal government claims that Americans enjoy "the safest food supply in the world," there is considerable evidence that our dinner plates contain healthy servings of some rather undesirable side dishes. The evidence has not escaped American consumers, who are increasingly questioning food safety. When the Food Marketing Institute, a trade association, asked consumers, "Is the food in your supermarket safe?" fully 75 percent weren't sure. Says the trade publication *Supermarket Business*: "The evidence suggests that food safety is poised to become one of the hottest public health issues of the 1990s."

Consider pesticides, for example. The statistics are truly mind-boggling:

- ❑ American farmers use 1.5 billion pounds of pesticides each year—about five pounds for every man, woman, and child. These pesticides end up in about half the foods we eat.
- ❑ Yet only about 1 percent of all food shipments are tested for pesticides. Indeed, standard government laboratory tests can't

even detect more than half of the 500 or so pesticides currently in agricultural use.

❑ According to the Environmental Protection Agency, 66 pesticides sprayed on food crops contain cancer-causing agents. Of the 560 million pounds of herbicides and fungicides used by American farmers annually, EPA says that 375 million pounds probably or possibly cause cancer.

❑ According to the National Academy of Sciences, only 10 percent of the 35,000 pesticides introduced since 1945 have been tested for their effects on people.

❑ Even worse, some 40 percent of pesticides are used to make food look good, according to a study by the California Public Interest Research Group.

❑ Some of the pesticides-in-food problem comes from outside U.S. borders: Some imported foods contain residues of dangerous pesticides that have been banned in the United States but are still used in other countries; one-fourth of all fresh fruit eaten by Americans comes from abroad.

❑ Perhaps most mind-boggling of all is that, despite this rampant use of pesticides, the National Academy of Sciences, considered the nation's preeminent body of scientists, reported in 1989 that farmers who apply little or no chemicals to crops are usually as productive as those who use pesticides and synthetic fertilizers.

Those particularly at risk from pesticides are children. According to a comprehensive two-year study released in 1989 by the Natural Resources Defense Council, pesticide tolerance levels established by the federal government are set only for healthy adults. Children face special danger: Because they weigh less they eat proportionally more pesticide-containing foods than adults. They also eat more fruit, which makes up an estimated one-third of preschoolers' diets, compared to one-fifth of adults'. Moreover, the younger a child is, the more susceptible he or she is to carcinogens. The NRDC study concluded that children will suffer significant additional cancer and brain damage from the pesticide residues in their food unless the Environmental Protection Agency tolerance levels are adjusted to account for children's consumption patterns and increased susceptibility.

In addition, the nitrogen-based fertilizers most commonly used in agriculture help to worsen the greenhouse effect, according to a study released in 1989 by the Marine Biological Laboratory in Woods Hole, Massachusetts, and the University of New Hampshire. The fertilizers, said the researchers, interfere with the ability of soil microbes to remove methane, one of the most potent greenhouse gases, from the air. The study pointed out that bacteria living in soil treated with nitrogen fertilizer take up far less methane gas, upon which they can feed, than do microbes in soil that has not been treated. Methane accounts for about one-fifth of the solar heat trapped by greenhouse gases. While carbon dioxide is far more concentrated in the atmosphere than methane, methane gas is twenty times more powerful at trapping heat close to the earth's surface. Methane levels have increased by about 1 percent a year over the past decade. (See "Seven Environmental Problems You Can Do Something About.")

And then there are the effects of pesticides on the water supply. Irrigation, which accounts for nearly 70 percent of groundwater use, can carry pesticides and fertilizers into rivers, lakes, and streams. It is estimated that up to 21 million pounds of pesticides reach groundwater or surface water before degrading. Environmental Protection Agency groundwater monitoring studies have found that at least 17 pesticides have been detected in groundwater in 23 states as a result of agricultural practices.

Pesticides and fertilizers aren't the only added ingredients affecting our food and our health. Another issue is agricultural drugs. Somewhere between 20,000 and 30,000 animal drugs are used in U.S. farming, including antibiotics, hormonal drugs, and other pharmaceuticals designed to create bigger, fatter, and faster-growing livestock. Residues of some of these drugs end up in milk, eggs, and meat.

Still another big problem is bacteria. According to the Centers for Disease Control, food-borne bacteria kill approximately 9,000 Americans each year. Millions of others suffer major or minor intestinal problems as a result of these germs. Much of this results from germs that grow in the intestines of animals that have become immune to germ-killing antibiotics. Another potent bacteria produces salmonella, which, according to the U.S. Department of Agriculture, may be present in as many as one out of every three chickens sold in the United States. But the Agriculture Department itself is part of the problem: In recent years, it has turned an increasing part of the food-inspection process over to the food industry itself. In effect, the inspectors are on company payrolls.

The sum total of all this has created an agricultural industry that is producing more and more food, but at a price. The price is paid for in the effect some of these practices have on our nation's farmland. The aggressive factory-farming techniques—which use drugs and pesticides as means of increasing production—have wreaked havoc on America's topsoil, the lifeblood of our agricultural economy, and on farmland in general. Indeed, some experts have said that today's high-tech farming methods will produce tomorrow's ecological disaster.

As the cost of farming has increased over the years, farmers have had to do more with less. They have had to increase crop yields and minimize crop losses with fewer people. They've relied more on machines, removing the hands-on aspect of farming, and on pesticides. Instead of planting a variety of crops, which causes less depletion of soil nutrients, they focus on a single crop, fueled by an increasing amount of artificial fertilizers. Most of these fertilizers are made in part from natural gas, a nonrenewable fossil fuel. It has been said that using these fertilizers is much like pouring a bunch of logs on a fire: the fire will burn brightly, then wither and die. So, too, with farming. The artificial fertilizers yield magnificent crops, but they also deplete the soil. And so, the following year, *more* fertilizers are needed, and the cycle continues. Conventional corn production results in the loss of an average of 20 tons of soil *per acre*.

In the end, the land is worked so hard with these techniques that it isn't able to replenish itself. Much of the soil is blown or washed away by rains faster than new soil is able to replenish it. Since the mid-1970s, U.S. farms have lost an average of 1.7 billion tons of soil every year. Our farmland is simply being swept away by unsustainable farming techniques.

**The New American Farmers.** A variety of less environmentally harmful farming methods are being employed by America's farmers. Not all of these methods are new; some have been around for decades. They go by a variety of names. "Low-impact farming," for example, involves a selective use of pesticides. "Crop rotation," which used to be standard practice, involves changing crop types regularly and discourages the reappearance of the same pests. And organic farming, which used to be an activity limited to

### *An Apple a Day . . .*

You may recall the Alar scare of 1989, when public concern arose over the use of the pesticide daminozide (marketed under the brand name Alar) on red apples. But Alar is only one of many problem chemicals, according to the U.S. Public Interest Research Group, a Washington, D.C.-based consumer group. USPIRG compiled a list of 23 other pesticides and agricultural chemicals used in apple farming that are suspected of containing cancer-causing substances:

benomyl
captafol
captan
chlordimeform
dicofol
dinoseb
folpet
lead arsenate
mancozeb
maneb
methomyl
metiram
o-phenylphenol
parathion
permethrin
phosmet
pronamide
silvez
simazine
tetrachlorvinphos
thiophanate-methyl
toxaphene
zineb

a relatively small group of "health-food hippies" dedicated to getting "back to the land," has entered the agricultural mainstream, as a growing number of farmers have learned how to produce healthy crop yields without synthetic fertilizers and pesticides. Other farming techniques have been known as "biodynamic farming," "biological farming," and "natural farming."

Whatever the name, the key word among today's forward-thinking farmers is "sustainability": the ability to produce crops without hurting the land. And the evidence shows that such farming techniques are good for both the land and its people. As the National Academy of Sciences reported in 1989, "Well-managed alternative farms use less synthetic chemical fertilizers, pesticides, and antibiotics without necessarily decreasing, and in some cases, increasing per-acre crop yields and the productivity of livestock systems." The academy recommended that the federal government revise its farm policy to encourage alternative farming methods that preserve the soil and reduce the use of chemicals.

Such statements have helped turn sustainable agriculture into more than simply a trendy alternative-lifestyles phenomenon. According to the **International Alliance for Sustainable Agriculture** (1701 University Ave. SE, Minneapolis, MN 55414; 612-331-1099), institutional researchers such as universities and food marketers have paid increasing attention—often in the form of research money—to organic farming methods. Several universities, including the University of California, Michigan State University, the University of Wisconsin, North Carolina State University, and Cornell University, have developed alternative agriculture research and extension programs.

Consumers are responding as well. According to a 1989 Harris Poll conducted by *Organic Gardening* magazine, when asked "Would you buy organically grown fruits and vegetables if they cost the same as other fruits and vegetables?" 84 percent of those polled answered yes. When that same 84 percent were asked "Would you still buy them if they cost more?" over half said yes.

## Organic Foods

Despite the various types of alternative farming, by the time produce reaches grocery shelves it is often labeled under the same name: organic. Unfortunately, "organic" is a much used and often abused term. Technically, *all* food is organic, in that its growth results from biological and chemical processes—even if those processes are stoked by tons of chemical fertilizers and pesticides. At the grocery store, however, the term "organic" has generally come to refer to foods grown without these chemicals.

While the precise definition of "organic" is still somewhat murky, some definitions have been accepted by a growing consensus. The following definition, for example, was agreed upon in 1989 by representatives of both organic and conventional agriculture in a meeting sponsored by the United Fresh Fruit and Vegetable Association, a trade group:

1. Organic food production systems are based on farm management practices that replenish and maintain soil fertility

by providing optimal conditions for soil biological activity.
2. Organic food is food that has been determined by an independent third party certification program to be produced in accordance with a nationally approved list of materials and practices.
3. Organic food is documented and verifiable by an accurate and comprehensive record of the production and handling system.
4. Only nationally approved materials have been used on the land and crops for at least three years prior to harvest.
5. Organic food has been grown, harvested, preserved, processed, stored, transported, and marketed in accordance with a nationally approved list of materials and practices.
6. Organic food meets all state and federal regulations governing the safety and quality of the food supply.

Is organic food really more healthful than nonorganic food? There is much debate on this. Skeptics note that an apple is an apple, no matter what coaxed it out of a tree. The nutritional value of the organic apple (or any other fruit or vegetable) will be virtually identical to the nonorganic one. Still, some nutritionists do claim that organically grown food has superior nutritional value.

Nutrition aside, it is the lack of toxic substances that may give organically grown food an advantage over most commercial produce. With the continuing sagas of contaminants in food—Alar in apples, aldicarb in watermelons, and EDB in grain, among others—organically grown produce, if not more nutritious, is less risky than other produce.

What's less risky for humans is probably less risky for the environment, at least in this case. When you buy organic foods, you are buying from farmers committed to the sustainability of the environment. And that's what being a Green Consumer is all about.

**Organic Certification.** So, how can you make sure that something truly is "organic"? A growing number of states—including California, Colorado, Iowa, Kansas, Maine, Minnesota, Montana, Nebraska, New Hampshire, New Mexico, North Dakota, Oregon, Texas, Vermont, Washington, and Wisconsin—have enacted organic labeling laws, most of which list the farming methods that can be used in order to be certified organic.

Another source of information is product labels certifying that a particular product meets certain criteria set by a professional organization. There are four principal organizations:

❑ **California Certified Organic Farmers** (Box 8136, Santa Cruz, CA 95061; 408-423-2263) calls itself "an organization of farmers and supporting members working to promote and verify organic farming practices, and supporting all efforts for a healthier, sustainable agricultural system."
❑ **Farm Verified Organic** (P.O. Box 45, Redding, CT 06875; 203-544-9896) describes itself as "an internationally accepted farm-to-table product-guarantee program that determines the authenticity of organically grown foods."
❑ **Organic Crop Improvement Association** (Box 729A, New Holland, PA 17557; 717-354-4936) describes itself as "an internationally recognized, brand-neutral, farmer-owned crop improvement, quality assurance program, which is backed by an independent third-party certification and audit-controlled system."
❑ **Organic Growers and Buyers Association** (1405 Silver Lake Rd., New Brighton, MN 55112; 612-636-7933) is a longtime organic certifier, which prides itself on "commitment to quality standards."

Each group has its own seal, which can appear only on products meeting each organization's standards. Those standards tend to be strict. All produce must be raised using acceptable products and techniques, and violators are subject to having their entire *field* disqualified for a number of years. Such requirements point up the longevity that many chemicals have in the soil; the life of these chemicals does not end when the produce is harvested.

In addition to these groups, many other growers set their own standards, considering themselves to be "self-certified." Some of these growers' standards are even more strict than the organizations' listed above; other growers' standards are less strict.

The **Organic Foods Production Association of North America** (OFPANA, P.O. Box 31, Belchertown, MA 01007; 413-323-6821) is a trade organization for the organic farming industry. Members include growers, processors, wholesalers, distributors, and retailers. OFPANA will supply its "Organic Farmers Associations Council" list of organic associations in the U.S. (send a self-addressed, stamped envelope). You may then locate the organizations in your area to determine who produces organic food locally, where it is sold, and what the standards are.

Another good source for information on organic programs, growers, and activities in your area is your state's agriculture department.

Still another organization actively involved in promoting healthy, organically grown food is **Americans for Safe Food** (ASF), a project of the Center for Science in the Public Interest (1501 16th St. NW, Washington, DC 2036; 202-332-9110), a respected consumer organization dealing primarily with health issues. ASF has attempted to organize consumers across the country to demand greater access to "contaminant-free" food at local grocery stores. The group publishes a list of reliable mail-order companies that will ship their organic products directly to individual consumers, usually via United Parcel Service; most do not require minimum orders.

The organizations listed below come from the ASF list. To obtain the most recent list, write to Americans for Safe Food at the address above; include $1 and a stamped, self-addressed legal-size envelope with 45 cents postage. To obtain price information, write or call each company for a complete product listing.

## Organic Food Mail-Order Suppliers

The following growers and distributors ship organic foods directly to individual consumers, usually via United Parcel Service, and do not require a minimum order unless otherwise noted. Write or call them for a complete product listing. Certification methods are indicated for each of the following suppliers: (C) indicates that all listed products have been certified organic by an independent agency; suppliers that establish their own standards are considered self-certified, labeled (S); (V) indicates that certification standards vary from product to product.

**Arizona:**
Arjoy Acres
HCR Box 1410
Payson, AZ 85541
602-474-1224
*Garlic, dried beans, peas. (S)*

**Arkansas:**
Dharma Farma
Star Route Box 140
Osage, AR 72638
501-553-2550
*Apples, pears. (C)*

Eagle Agricultural Products
407 Church Ave.
Huntsville, AR 72740
501-738-2203
*Fresh and dried produce, beans, grains, pasta, flour. (V)*

Good Earth Association
202 E. Church St.
Pocahontas, AR 72455
501-892-9545, 501-892-8329
*Corn, beans, seed crops. (S)*

Mountain Ark Trading Co.
120 S. East Ave.
Fayetteville, AR 72701
501-442-7191, 800-643-8909
*Grains, beans, seeds, wide selection of products. (V)*

**California:**
Ahler's Organic Date Garden
P.O. Box 726
Mecca, CA 92254
619-396-2337
*Dates, date products. (S)*

Blue Heron Farm
P.O. Box 68
Rumsey, CA 95679
916-796-3799
*Almonds, walnuts, oranges. Minimum: 5-35 pounds depending on item. (C)*

Covalda Date Co.
P.O. Box 908
Coachella, CA 92236
619-398-3441
*Dates, dried fruits and nuts. (S)*

Capay Fruits and Vegetables
Star Route, Box 3
Capay, CA 95607
916-796-4111
*Dried tomatoes, peaches, herbs. (C)*

Dach Ranch
P.O. Box 44
Philo, CA 95466
707-895-3173
*Apples, apple products, pears. (C)*

Ecology Sound Farms
42126 Rd. 168
Orosi, CA 93647
209-528-3816, 528-2276
*Oranges, plums, Asian pears, kiwifruit, persimmons. Minimum: 8-40 pounds. depending on item. (C)*

Frey Vineyards
14000 Tomki Rd.
Redwood Valley, CA 95470
707-485-5177
*Wine. Minimum: one-half case. (C)*

Gold Mine Natural Food Co.
1947 30th St.
San Diego, CA 92102
619-234-9711, 800-647-2929
*Wide variety of goods. (V)*

Gravelly Ridge Farms
Star Route 16
Elk Creek, CA 95939
916-963-3216
*Produce, grains. (C)*

Great Date in the Morning
P.O. Box 31
Coachella, CA 92236
619-398-6171
*Dates. (C)*

Green Knoll Farm
P.O. Box 434
Gridley, CA 95948
916-846-3431
Kiwifruit. Minimum: 8 pounds. (C)

Jaffe Brothers
P.O. Box 636
Valley Center, CA 92082
619-749-1133
*Dried fruits, nuts, grains, beans, assorted goods. Minimum: 5 pounds. (S)*

Living Tree Centre
P.O. Box 797
Bolinas, CA 94924
415-868-2224
*Almonds, almond butter, apples. (C)*

Lundberg Family Farm
P.O. Box 369
Richvale, CA 95974
916-882-4551
*Rice and rice products. (C)*

Mendocino Sea Vegetable Co.
P.O. Box 372
Navarro, CA 95463
707-895-3741
*Wildcrafted sea vegetables. (S)*

Old Mill Farm School of Country Living
P.O. Box 463
Mendoncino, CA 95460
707-937-0244
*Lamb, goat cheese, produce. Minimum: $50. (C)*

Steven Pavich and Sons
Route 2, Box 291
Delano, CA 93215
805-725-1046
*Grapes. (C)*

Sleepy Hollow Farm
44001 Dunlap Rd.
Miramonte, CA 93641
209-336-2444
*Apples, cooking herbs. Minimum: one box. (C)*

Joe Soghomonian
8624 S. Chestnut
Fresno, CA 93725
209-834-2772
*Grapes in season, raisins. (C)*

Timber Crest Farms
4791 Dry Creek Rd.
Healdsburg, CA 95448
707-433-8251
*Wide variety of dried fruits, nuts. (S)*

Weiss' Kiwifruit
594 Paseo Compañeros
Chico, CA 95928
916-343-2354
*Kiwifruit. Minimum: two and a half pounds. (C)*

Your Land Our Land
P.O. Box 485
Los Altos, CA 94022
415-821-6732
*Fresh produce, herbs, garlic. (V)*

**Colorado:**
Malachite Small Farm School
ASR Box 21- Pass Creek Rd.
Gardner, CO 81040
719-746-2412
*Honey, quinoa, beef. (S)*

Wilton's Organic Potatoes
Box 28
Aspen, CO 81612
303-925-3433
*Potatoes. Minimum: 5 pounds. (C)*

**Connecticut:**
Butterbrooke Farm
78 Barry Rd.
Oxford, CT 06483
203-888-2000
*Vegetable seeds. (C)*

**District of Columbia:**
Tabard Farm Potato Chips
1739 N St. NW
Washington, DC 20036
202-785-1277
*Potato chips. (S)*

**Florida:**
Sprout Delights
13090 NW 7th Ave.
Miami, FL 33168
305-687-5880, 800-334-2253
*Wide range of akery items. Minimum: $20. (S)*

Starr Organic Produce Inc.
P.O. Box 561502
Miami, FL 33256
305-262-1242
*Wide variety of fruit. Minimum: 20 pounds. (S)*

**Hawaii:**
Hawaiian Exotic Fruit Co.
Box 1729
Pahoa, HI 96778
808-965-7154
*Dried fruit, fresh ginger root, tumeric. Minimum: 10 pounds. (S)*

**Idaho:**
Famous Idaho Potatoes
Star Route
Moyie Springs, ID 83845
208-267-7938
*Vegetables. Minimum: $10. (S)*

**Illinois:**
Green Earth Natural Foods
2545 Prairie Ave.
Evanston, IL 60201
312-864-8949, 800-322-3662
*Wide variety of items. (V)*

Nu-World Amaranth
P.O. Box 2202
Naperville, IL 60565
312-369-6819
*Amaranth flour, cereal, whole grain. (C)*

Sunrise Farm Health Food Store
17650 Torrence Ave.
Lansing, IL 60438
312-474-6166
*Wide variety of items. (V)*

## Rain Forest Nuts

Some nuts from South American rain forests are sure-fire "green" shopping buys. The farming and harvesting of such nuts as cashews and Brazil nuts support rain forest preservation. Brazil nuts, for example, are grown only in rain forests and cannot be grown on plantations. The nuts provide people who live in rain forests with an economic alternative to cutting the forests down, allowing them to harvest the rain forests' products in a way that retains the forest rather than destroying it. Cashews are also a green buy, because they are often grown on previously degraded lands and are frequently used in reforestation plans. Buying cashews encourages reclamation of rain forest lands. Not only does nut farming promote forest conservation, it is five times as profitable per acre as cattle ranching, according to Cultural Survival, a Massachusetts consulting organization.

Supermarket shelves are a good place to shop for rain forest nuts. In addition, there is a growing number of products being made with rain forest nuts, including Rainforest Crunch, a nut brittle made with nuts harvested from Brazilian rain forests. Eight-ounce boxes of Rainforest

**Indiana:**
Dutch Mill Cheese
2001 N. State Rd. I
Cambridge City, IN 47327
317-478-5847
*Cheese. Minimum: 3 pounds. (S)*

**Iowa:**
Frontier Cooperative Herbs
P.O. Box 299
Norway, IA 52318
319-227-7991
*Herbs, spices, teas. (V)*

Paul's Grains
2475-B 340 St.
Laurel, IA 50141
515-476-3373
*Whole grains and grain products, beef, lamb, chicken, turkey. (S)*

**Kentucky:**
Gracious Living Farm
General Delivery
Insko, KY 41443
*Vegetables. (S)*

**Louisiana:**
Rein Farms
812 Cedar Ave.
Metairie, LA 70001
504-888-5763
*Vegetables. (S)*

**Maine:**
Crossroad Farms
Box 3230
Jonesport, ME 04649
207-497-2641
*Root crops, squash, cabbage, apples. Minimum: $25. (C)*

Crunch are available in **Ben & Jerry's** ice-cream stores, gourmet and specialty stores, some convenience stores (such as Food and Liquor in northern California), and **L. L. Bean**and **Patagonia** retail stores. (Ben & Jerry's also offers Rainforest Crunch ice cream in its retail "scoop shops," although not yet in retail stores.) Also, two-box packs and one-pound tins are available through the mail directly from **Community Products** (RD #2, Box 1950, Montpelier, VT 05602; 802-229-1840). Mail-order prices are $8.80 for orders east of Chicago; $9.60 for orders in and west of Chicago. Forty percent of the profits from Rainforest Crunch is donated to organizations working to protect rain forests; an additional 20 percent of profits is given to the "1% for Peace" fund.

Another organization, **Cultural Survival Imports** (11 Divinity Ave., Cambridge, MA 02138; 617-495-2562), imports sustainably managed rain forest nuts bought from the harvesters themselves. The nuts are available directly from Cultural Survival; call for price lists and order forms. Groups can also buy the nuts to sell as a fundraiser. At least 40 percent of Cultural Survival's profits go to various groups working to protect rain forests.

Fiddler's Green Farm
R.R. 1, Box 656
Belfast, ME 04915
207-338-3568
*Numerous whole grain mixes, coffee, syrup, jam. (V)*

Johnny's Selected Seeds
Foss Hill Rd.
Albion, ME 04910
207-437-9294
*Vegetable, herb, and farm seeds. (S)*

Maine Coast Sea Vegetables
Shore Rd.
Franklin, ME 04634
207-565-2907
*Sea chips, kelp, dulse, nori. (S)*

Simply Pure Food
RFD #3, Box 99
Bangor, ME 04401
207-941-1924, 800-426-7873
*Strained and diced baby foods, baby cereals. (C)*

**Maryland:**
Macrobiotic Mall
18779-C N. Frederick Ave.
Gaithersburg, MD 20879
301-963-9235, 800-533-1270
*Grains, legumes, packaged foods, macrobiotic items. (C)*

Organic Foods Express
11003 Emack Rd.
Beltsville, MD 20705
301-937-8608
*Produce, grains, beans, coffee. (V)*

**Massachusetts:**
Baldwin Hill Bakery
Baldwin Hill Rd.
Phillipston, MA 01331
508-249-4691
*Sourdough bread. Minimum: 12 loaves. (C)*

Cooks Maple Products
Bashan Hill Rd.
Worthington, MA 01098
413-238-5827
*Maple syrup. (S)*

Greek Gourmet Ltd.
195 Whiting St.
Hingham, MA 02043
617-749-1866
*Extra virgin olive oil, olives. Minimum: 1 case. (S)*

**Michigan:**
Cosmic Realities
P.O. Box 1250
Jackson, MI 49204
517-783-2293
*Sprouted foods, grains, beans. (C & S)*

Country Life Natural Foods
109th Ave.
Pullman, MI 49450
616-236-5011
*Bulk natural foods. (V)*

Eugene and Joan Saintz
2225 63rd St.
Fennville, MI 49408
616-561-2761
*Fresh produce in season. (S)*

Specialty Grain Co.
Box 2458
Dearborn, MI 48123
313-535-9222
*Grains, beans, seeds, dried fruits, nuts. (C)*

**Minnesota:**
Diamond K Enterprises
R.R. 1, Box 30A
St. Charles, MN 55972
507-932-4308, 932-5433
*Grains and grain products, nuts, dried fruits. (C)*

French Meadow Bakery
2610 Lyndale Ave. S.
Minneapolis, MN 55408
612-870-4740
*Many sourdough breads. Minimum: $20. (C)*

Living Farms
Box 50
Tracy, MN 55408
800-622-5235 in state, 800-533-5320 out of state.
*Grains, sprouting seeds. (V)*

Mill City Sourdough Bakery
1566 Randolph Ave.
St. Paul, MN 55105
612-698-4705, 800-873-6844
*Sourdough breads. Minimum: 6 loaves. (C)*

Natural Way Mills, Inc.
Route 2, Box 37
Middle River, MN 56737
218-222-3677
*Whole grains, flours, cereals, other products. (C)*

**Missouri:**
Morningland Dairy
Route 1, Box 188-B
Mountain View, MO 65548
417-469-3817
*Raw milk cheeses. (S)*

**Nebraska:**
Do-R-Dye Organic Mill
Box 50
Rosalie, NE 68055
402-863-2248
*Oats, wheat, rye, corn products. Minimum: $5. (S)*

Stapelman Meats
Route 2, Box 6A
Belden, NE 68717
402-985-2470
*Beef, pork. (S)*

**New Hampshire:**
Water Wheel Sugar House
Route 2
Jefferson, NH 03583
603-586-4479
*Maple syrup. (S)*

**New Jersey:**
Simply Delicious
243 A N. Hook Rd., Box 124
Pennsville, NJ 08070
609-678-4488
*Wide variety of items. (V)*

**New Mexico:**
Estawanca Valley Garlic Growers
P.O. Box 892
Moriarty, NM 87035
505-832-6177
*Elephant garlic. Minimum: $10. (S)*

**New York:**
Bread Alone
Route 28
Boiceville, NY 12412
914-657-3328
*Wheat, rye, sourdough breads. Minimum: 12 loaves. (C)*

Community Mill and Bean
RD 1, Route 89
Savannah, NY 13146
315-365-2664
*Flour, mixes, beans, grains, cereals. Minimum: $10. (C)*

Deer Valley Farm
RD 1
Guilford, NY 13780
607-764-8556
*Meats, produce, grains, baked goods, wide variety of products. Minimum: $10. (V)*

Four Chimneys Farm Winery
RD 1, Hall Rd.
Himrod, NY 14842
607-243-7502
*Wine, grape juice, wine vinegar. Minimum: juice, case only. (C)*

**North Carolina:**
American Forest Foods Corp.
Route 5, Box 84E
Henderson, NC 27536
919-438-2674
*Shiitake and oyster mushrooms, mixes, spices. Minimum: 12 packages. (C)*

**North Dakota:**
Hugh's Gardens
Route 1, Box 67
Buxton, ND 58218
701-942-3345
*Carrots, onions, potatoes. Minimum: $50. (C)*

**Ohio:**

Millstream Natural Health Supplies
1310A E. Tallmadge Ave.
Akron, OH 44310
216-630-2700
*Produce, nuts, grains, cereals. (V)*

Sanctuary Farm
RD 1, Butler Rd., Box 184A
New London, OH 44851
419-929-8177
*Grains, beans, seeds. Minimum: 25 pounds. (C)*

**Oregon:**

Dement Creek Farms
Box 155
Broadbent, OR 97414
503-572-5564
*Beans, grains, vegetables, herbs. (C)*

Herb Pharm
P.O. Box 116
Williams, OR 97544
503-846-7178
*Herbs, herbal extracts, teas. Minimum: $25. (C)*

**Pennsylvania:**

Dutch Country Gardens
Box 1122, RD 1
Tamaqua, PA 18252
717-668-0441
*Potatoes, carrots. Minimum: $25. (S)*

Garden Spot Distributors
438 White Oak Rd.
New Holland, PA 17557
717-354-4936
*Wide variety of dried and packaged goods. (V)*

Genesee Natural Foods
RD 2, Box 105
Genesee, PA 16923
814-228-3200, 228-3205
*Beans, grains, flours, honey, pasta, cereals, dried fruit. Minimum: $20. (V)*

Krystal Wharf Farms
RD 2, Box 191A
Mansfield, PA 16933
717-549-8194
*Grains, beans, nuts, dried fruit, fresh produce and other products. (C & S)*

Rising Sun Distributors
P.O. Box 627
Milesburg, PA 16853
814-355-9850
*Produce, dried fruit, nuts, beans, grains, other items. (V)*

Walnut Acres
Penns Creek Rd.
Penns Creek, PA 17862
717-837-0601
*Full line of cerelas, flours, grains, baked goods, vegetables, other items. (S)*

**Texas:**

Arrowhead Mills
Box 2059
Hereford, TX 79045
806-364-0730
*Wide variety of grains and grocery items. (C)*

J. Francis Co.
Route 3, Box 54
Atlanta, TX 75551
214-796-5364
*Pecans. Minimum: 5 pounds. (S)*

Stanley Jacobson
1505 Doherty
Mission, TX 78572
512-585-1712
*Grapefruit and oranges. Minimum: one-fourth bushel. (C)*

Lee's Organic Foods
Box 111
Wellington, TX 79095
806-447-5445
*Fruit jerkies. Minimum: $5. (C)*

**Utah:**
Aquaculture Marketing Service
356 W. Redview Dr.
Monroe, UT 84754
800-634-5463, ext. 230
*Frozen, canned and smoked rainbow trout. (S)*

**Vermont:**
Hill and Dale Farms
West Hill-Daniel Davis Rd.
Putney, VT 05346
802-387-5817
*Apples, vinegar. Minimum: one flat of 24 apples. (S)*

Teago Hill Farm
Barber Hill Rd.
Pomfret, VT 05067
802-457-3507
*Maple syrup. (S)*

**Virginia:**
Blue Ridge Food Service
Route 3, Box 304
Edinburg, VA 22824
703-459-3379
*Rainbow trout. Minimum: 5 pounds. (S)*

Golden Acres Orchard
Route 2, Box 2450
Front Royal, VA 22630
703-636-9611
*Apples in season, cider, vinegar, juice. (S)*

Golden Angels Apiary
P.O. Box 2
Singers Glen, VA 22850
703-833-5104
*Five types of honey. (S)*

Kennedy's Natural Foods
1051 W. Broad St.
Falls Church, VA 22046
703-533-8484
*Wide variety of items. (V)*

Natural Beef Farms
4399-A Henninger Ct.
Chantilly, VA 22021
703-631-0881
*Frozen meats, produce, bread, wide variety of items. (C)*

**Washington:**
Cascadian Farm
5375 Highway 20
Rockport, WA 98283
206-853-8175
*Fruit conserves, dill pickles. (C)*

Homestead Organic Produce
Route 1, 2002 RD 7 NW
Quincy, WA 98848
509-787-2248
*Onions, garlic, apples, nuts, dried fruits. Minimum: varies. (C)*

**West Virginia:**
Brier Run Farm
Route 1, Box 73
Birch River, WV 26610
304-649-2975
*Goat cheeses. (S)*

Hardscrabble Enterprises Inc.
Route 6, Box 42
Cherry Grove, WV 26804
304-567-2727, 202-332-0232
*Dried shiitake mushrooms. Minimum: one and a half pounds. (S)*

**Wisconsin:**
Joel Afdahl
Route 1, Box 270
Hammond, WI 54015
715-796-5395
*Maple syrup. (S)*

Nokomis Farm
3293 Main St.
East Troy, WI 53120
414-642-9665
*Grains, breads, beef, pork. (C)*

## Organic Baby Foods

If you are at all concerned about the potential dangers in large-scale fertilizer-and-pesticide farming, consider your young child: with his or her mind and body developing so quickly, nearly everything consumed can have an effect. In young children, cells are rapidly dividing, and organs, like the liver, may not be as efficient in removing toxic chemicals. As stated earlier, the younger a child is, the more susceptible he or she will be to carcinogens. According to the Natural Resources Defense Council, between 5,500 and 6,200 of today's preschoolers are likely to develop cancer over their lifetimes solely because of exposure to just eight pesticides. Some of the pesticide scares of recent years have spurred parents toward organic baby foods; other parents have started making their own foods, something that nearly every mother did only a generation or so ago.

There are two principal manufacturers of organic baby food:

❑ **Earth's Best Baby Food** (P.O. Box 887, Middlebury, VT 05753; 802-388-7974, 800-442-4221) is available in natural food stores nationwide, in supermarkets in the Northeast and in Colorado, and by mail order in case quantities from **Hand In Hand** (9180 LeSaint Dr., Fairfield, OH 45014; 513-874-1665, 800-543-4343). It is also available through diaper services in several cities.

❑ **Simply Pure Baby Food** (RFD #3, Box 99, Bangor, ME 04401; 207-941-1924, 800-426-7873) is available in some natural food stores in the Northeast, but is sold mostly by mail order.

To obtain price information, write or call each company for a complete product listing.

A word about price: Unfortunately, organically grown baby food can cost up to three times as much as other brands.

## The Safety of Seafood

Seafood is generally considered the most healthful, nutritious, low-fat alternative to red meat and chicken. The recent warning from the Surgeon General and the National Academy of Sciences that the typical American diet is too high in fat, thereby increasing risk of heart disease and other health problems, has spurred Americans' consumption of fish and shellfish. A combination of two major problems is threatening the otherwise good reputation of seafood, however: contaminated waters from which seafood is caught, and a lack of comprehensive, mandatory federal inspections of seafood processing facilities.

As we pointed out earlier, a variety of toxic substances persist in the environment and are spread to our water sources. (See "Seven Environmental Problems You Can Do Something About.") Municipalities and industries dump their sewage and other wastes directly into our rivers and oceans. Waters from which fish are taken may be polluted with industrial chemicals and agricultural run-off such as pesticides.

Unlike land-based food products, fish and shellfish are generally harvested from uncontrolled environments in international waters. Moreover, some types of seafood actually concentrate contaminants in their fatty tissue. In other words, they act like sponges when absorbing water-borne contaminants. An individual fish can contain more toxins than the water in which it is caught.

In addition, the seafood supply can lead to serious health risks due to the inadequacy of current federal and state inspection programs. Currently, seafood is the only flesh food in this country not subject to a comprehensive, mandatory inspection program. **Public Voice for Food and Health Policy**, a consumer organization, has called for a federal inspection program that would establish federal microbiological and chemical standards; improve import programs for contaminants; establish "traceback" (the ability to trace a contaminant to its source) and record-keeping systems for fish, especially shellfish, of their sources of origin; and develop testing methods for toxins. It also calls for mandatory improved consumer information programs. For more information on these efforts for compulsory seafood inspection contact **Public Voice for Food and Health Policy**, 1001 Connecticut Ave. NW, Suite 522, Washington, DC 20036; 202-659-5930.

**Tuna and Dolphins.** It's unfortunate when a relatively healthy, renewable fresh food gets enmeshed in a controversy that makes it an undesirable purchase for Green Consumers. That's exactly the case with canned tuna.

The problem has nothing to do with the tuna itself. Indeed, tuna, like most fish, represents a low-fat, high-protein food that has rarely been criticized by nutritionists. Environmentalists, however, have a different view. In the process of fishing for tuna, international fishing fleets in the eastern tropical Pacific Ocean slaughter as many as 150,000 dolphins a year, according to the Earth Island Institute Dolphin Project.

The reason: Yellowfin tuna, for reasons unknown, often gather just below herds of dolphin. So tuna fleets watch for dolphins to locate their catch. Then, in a practice known as "setting on dolphins," fishermen chase the dolphins with helicopters and speedboats. Exhausted and terrified, the dolphins are then encircled in nets up to a mile long that are drawn closed at the bottom. Trapped, the air-breathing dolphins suffocate or drown.

The result: Mixed in with a hundred dead or dying dolphins may be a dozen or more tuna. The dead or wounded dolphins are cast back into the sea.

As the Earth Island Institute puts it: "Dolphins aren't fish. They're mammals like us. They breathe air. They nurse their young. They communicate and work together in groups. Graceful and intelligent creatures, dolphins have been known to come to the aid of drowning humans."

The Marine Mammal Protection Act, passed in 1972, was intended to eliminate such dolphin deaths. Progress in reducing the deaths was made until 1981, when the Reagan administration relaxed the dolphin quotas. But the biggest problem comes from foreign fishing fleets, now dominating tuna fishing, which have no restrictions at all on killing dolphins.

In addition to seeking laws to prohibit dolphin setting and to embargo tuna from countries that do not comply with marine mammal protection laws, environmental groups like Earth Island Institute, Greenpeace, and the Humane Society of the United States have called on consumers to boycott tuna. *All brands of tuna are suspect,* including albacore and bonita—chunk, solid, and flaked, packed in water or oil.

Keep this in mind: Every day that these tuna-fishing practices continue, up to 1,000 more wild dolphins will be trapped and drowned in the nets.

For more information, contact one of the following environmental organizations:

Earth Island Institute
300 Broadway, Suite 28
San Francisco, CA 94133
415-788-3666

Humane Society of the U.S.
2100 L St. NW
Washington, DC 20037
202-452-1100

Greenpeace
P.O. Box 3720
Washington, DC 20007
202-462-1177

## Fast Food and the Environment

Whether or not to eat fast food is largely a nutritional question, of course. But we're not going to deal with that here. We're only going to examine what fast-food companies are doing to the environment.

There are two principal areas of concern:

❑ **Hamburgers and the Rain Forests.** Only about 2 percent of the beef used in fast-food hamburgers comes from rain forests in Central America. While a seemingly insignificant amount, the process is having a devastating effect on the rain forests and on the disappearing forests' contribution to the greenhouse effect, among other environmental problems. (See "Seven Environmental Problems You Can Do Something About.")

According to the Rainforest Action Network (RAN), "The typical four-ounce hamburger patty represents about 55 square feet of tropical forest—a space that would statistically contain one 60-foot-tall tree; 50 saplings and seedlings representing 20 to 30 different tree species; two pounds of insects representing thousands of individuals and more than a hundred different species; a pound of mosses, fungi, and micro-organisms; and a section of the feeding zone of dozens of birds, reptiles, and mammals, some of them extremely rare. Millions of individuals and thousands of species of plants and animals inhabit a patch of tropical forest destroyed for a single hamburger."

Who uses the beef from rain forests? You won't find much of it in the supermarket as ground beef and steak; beef raised in rain forest regions is said to be stringy, tough, and cheap and often goes into mass-produced, processed foods, where it is combined with fattier domestic beef and cereal products. Although hamburger chains like **McDonald's** had been accused of causing tropical deforestation, it is now clear that McDonald's local raw materials sourcing policy has meant that the company takes vigorous steps to ensure that its business does not damage tropical rain forests. The company's policy in this respect, announced in 1989, stresses that

> . . . it is McDonald's policy to use only locally produced and processed beef in every country where we have restaurants. In those isolated areas where domestic beef is not available, it is imported from approved suppliers in other countries. In all cases, however, McDonald's does not, has not, and will not permit the destruction of tropical rain forests for our beef supply. We do not, have not, and will not purchase beef from rain forests or recently deforested rain forest land. This policy is strictly enforced and closely monitored. Any McDonald's supplier who is found to deviate from this policy or who cannot prove compliance with it will be immediately discontinued.

Environmentalists have had some difficulty determining exactly which of the big fast-food hamburger chains use rain forest beef. A national boycott of **Burger King**, part of the Pillsbury Company, called by Earth First! and other environmental groups, was called off in 1989 after Burger King announced it would no longer buy Central American beef. (However, the company still had not provided verification as this book went to press.) Other companies such as **Wendy's** and **Roy Rogers** deny they use rain forest beef. These denials notwithstanding, Central American rain forests contribute about 130 million pounds of meat to America's beef eaters each year.

Fast-food hamburgers aren't the only destination for rain forest beef, however. Rainforest Action Network has urged Americans to avoid purchasing such processed-beef products as:

❑ baby foods
❑ canned beef products
❑ frozen beef products
❑ hot dogs
❑ luncheon meats
❑ soups

Another company being boycotted by Earth First! is the **Campbell Soup Company**, which has stated that while it does import beef from South America, none of the beef comes from rain forest areas. According to RAN, "There is another reason to avoid beef imported from Latin America: this beef is more likely to be contaminated with toxic chemicals, trace metals, and organic contaminants than that raised in the United States. Excessive pesticide residues have repeatedly been found in beef prepared by packing plants in Costa Rica, El Salvador, Guatemala, and Mexico, for which these packers have been decertified repeatedly by the U.S. Department of Agriculture."

❑ **Plastic (and Other) Trash.** We've already discussed the problem of solid waste (see "Seven Environmental Problems You Can Do Something About") and the mountain of trash created, among other things, by plastic (see "The Problem of Packaging"). It comes as no surprise, then, that much of that plastic—including a lot of polystyrene foam packaging—is used by the fast-food industry.

Environmental groups have focused on McDonald's in their fight to ban the use of foam containers by fast-food companies. According to the Citizen's Clearinghouse for Hazardous Waste, McDonald's is believed to generate 1.6 billion cubic feet of "Styrotrash" in the form of polystyrene "clamshell" hamburger containers. (Besides packaging Big Macs and Quarter Pounders, these containers were specifically promoted in McDonald's "hot side hot, cool side cool" ads promoting its McDLT. Keeping various parts of McDLTs hot and cool required creating a polystyrene container nearly twice the size of regular hamburger containers.) According to another environmental group, a strip of 55 billion end-to-end burger containers—the number of burgers McDonald's claims to have served—would circle the earth 333 times at the equator.

McDonald's met its environmental critics headon with its 1987 announcement that stated the company's intention to discontinue the use of foam packaging made with chlorofluorocarbons, switching instead to polystyrene containers made with CFC-free hydrocarbon-based blowing agents. But environmentalists have countered that, while lacking CFCs, such packaging still does not biodegrade, contributing to landfill problems, and is likely to give off toxic fumes when burned in incinerators.

McDonald's and many other large and small eateries usually point to cost when defending their decisions to use foam containers. But according to a study in Washington state, paper cups with a wax coating for cold liquids sold for about $ .033 per cup, while foam was about $ .018 per cup—a difference of one and a half cents. (The figures assume one is purchasing cups in large quantities.) Cups for hot liquids were a bit more expensive because of the amount of paper necessary for insulation—$ .039 per cup compared to $ .013 for polystyrene foam cups. Paper plates were found to be about a penny less expensive than polystyrene foam. Such figures seem minuscule, given the hundreds of millions some fast-food companies spend on advertising alone each year. Besides, surveys indicate that most fast-food consumers would likely be willing to pay two cents extra per item if they knew that the extra money would go to pay for more environmentally responsible packaging materials.

The extra costs aside, some major fast-food outlets have made the switch to paper. In 1988, Wendy's replaced its foam sandwich containers with paper and foil wrappers. (Wendy's also replaced its coffee cups and disposable plates with non-CFC foam packaging.) Burger King uses cardboard instead of plastic for its Whopper and other sandwiches.

As with automobiles, there is no such thing as a green fast-food restaurant. As with automobiles, most fast food is assembled from parts shipped in from various locations, some outside the country—the beef from one state, the potatoes from another, the oil, chicken, fish, cups, plates, and paper bags from still others. By virtue of the energy costs needed to haul all these products and materials to the tens of thousands of fast-food outlets around the world, these products are far from green. And the vast amount of packaging—whether foam, plastic, cardboard, or paper—still makes a generous contribution to our nation's trash heap.

Your greenest choice is to avoid fast food whenever possible. It isn't healthy—for you or for the planet.

## Household Paper Products

Paper. What could be more environmentally responsible? It comes from natural, renewable resources and degrades easily and quickly when thrown away. How bad could a paper towel be, anyway?

Plenty bad. In fact, it could be downright deadly.

The problem with paper towels—and many other household paper products, including toilet paper, disposable diapers, coffee filters, milk cartons, tampons, and facial tissues—has to do with the presence of dioxins, a family of toxic, carcinogenic chemicals, one of which was used as a weapon by American troops in Vietnam as the deadly "Agent Orange." A growing body of research has found traces of dioxins in many consumer paper products. There is considerable concern about the health effects resulting when these products come into contact with food or with sensitive parts of the body.

What are dioxins and how do they get into paper? "Dioxin" refers to a chemical family with 75 individual members. While all have the same basic chemical structure, some are more toxic than others. Scientists believe that dioxins imitate natural steroid hormones, such as estrogen, in our bodies and can trigger a wide range of biochemical reactions. Minute quantities can trigger anything from acne and achy joints to insomnia, cancer, birth defects, and immune system disorders. Moreover, dioxins (and a chemical cousin called "furans") tend to accumulate in the body because they are fat soluble; they are stored in the fat cells of each organism. As always, children are especially sensitive—dioxin is found in the milk of the average North American mother. It is possible for nursing infants to be exposed to up to two hundred times more dioxin than healthy adults. Based on animal tests, the Environmental Protection Agency has classified dioxins as a "probable human carcinogen."

Dioxins are formed in pulp making during the chlorine bleaching process. The problem is most severe in the manufacturing of bleached "kraft" pulp—a process that uses a sulfide soak to cook the wood chips in. The process produces a strong, dark-colored pulp suitable as feedstock for manufacturing paper. It must then be bleached in a five- or six-stage sequence to achieve high brightness. Chlorine is used to bleach the pulp. Higher-grade papers, such as most printing papers, require fewer bleaching stages and less chlorine, resulting in less dioxin formation. Newsprint, because it is not bleached, is not known to contain dioxins.

The dioxins created in pulp making don't just end up in the paper. In fact, scientists first made the link between dioxin and paper after discovering unexpectedly high levels of the chemicals in several fish downstream of pulp and paper mills. The high dioxin levels in the fish made them unsafe for human consumption. Moreover, when dioxin-contaminated water is used to irrigate crops, they, too, become contaminated. The Environmental Protection Agency reported in 1988 that, when grown in contaminated soils, root crops—such as carrots, potatoes, and onions—can develop dioxin levels that equal or exceed those found in the soil itself.

But it is the paper products themselves that present the most immediate problem. An official at the Food and Drug Administration told *Science News* in 1989 that the two major sources of dioxins were milk cartons and coffee filters. The official's "very rough estimates" were that young children getting all their milk from contaminated cartons might double their dioxin intake. Heavy coffee drinkers consuming most of their brew from pots with bleached-paper filters might increase their daily dioxin intake 5 to 10 percent above the average U.S. level.

But almost all bleached paper products made from virgin pulp are suspect. Tests sponsored by the paper industry found that

## *Recycled Paper Products*

In addition to being concerned about bleaching, Green Consumers should look for paper products that are recycled as well. Some of the companies listed on pages 112-113 offer recycled as well as unbleached paper. Sources include **Marcal Paper Mills, Inc.** (Market St., Elmwood Park, NJ 07407; 201-796-4000, 800-631-8451), **Ashdun Industries** (1605 John St., Fort Lee, NJ 07024; 201-944-2650), and **Ft. Howard Corporation** (P.O. Box 19130, Green Bay, WI 54307; 414-435-8821), which makes toilet paper, paper towels, and other products from recycled fiber.

See also the "Green Gifts" chapter for other companies selling recycled paper products.

superabsorbent disposable diapers, paper towels, tea bags, tampons, juice cartons, TV dinner containers, and various types of paper plates also have contained low levels of dioxin. Dioxins from any of these products can migrate into foods or onto sensitive parts of bodies. Moreover, when these paper products are disposed of in a landfill or are incinerated, the dioxins released in the air can be inhaled by animals and humans or ingested through contamination of food crops.

It is important to keep in mind that *the smallest detectable amounts of these compounds have been known to cause cancer in laboratory animals.*

By the way, dioxins aren't the only toxic substances found in the wastewater of pulp and paper mills. A 1986 study by the Ontario Ministry of the Environment found a total of 41 substances of concern, from aluminum to zinc. Included in that list were some of the most notorious pollutants: benzene, cadmium, lead, mercury, polychlorinated biphenyls (PCBs), toluene, and others. Ironically, some environmentalists fear that focusing public attention on dioxins might derail industry and government action on some of these other pollutants.

**Avoiding Dioxins.** The good news is that you can avoid the hazards of dioxins in paper products by buying unbleached or chlorine-free paper products. The bad news is that, due to lower

## *Paper Bags or Plastic?*

One of the great debates among Green Consumers is the following: At the grocery, is it better to carry your purchases home in paper bags or plastic bags?

The answer seems simple enough. Paper is a renewable resource that biodegrades easily. Plastic is made from nonrenewable petroleum products and can be only partially recycled with difficulty; more than likely, it will take decades or centuries to biodegrade.

But the world, as we well know, is not that simple. Paper bags also present environmental problems. In fact, the paper-versus-plastic question is a no-win proposition.

The problem with brown paper bags is that they are made entirely from virgin paper; bags made of 100 percent recycled paper simply do not provide the strength needed to hold the weight of a full bag of groceries without breaking. As strange as it may seem, the use of virgin paper can be more damaging to the environment than some plastic substitutes. The amount of sulfur dioxide, nitrogen oxides, carbon monoxide, and dust are greater during the manufacture of paper bags than during that of polyethylene plastic bags. Wastewater pollutants are also many times higher during the production of paper bags.

None of this, of course, is meant to endorse plastic bags. Far from it. They are polluting too, even the so-called "biodegradable" and "recyclable" ones. (See "The Perils of Plastic" for more on this.)

In choosing between paper and plastic, it is important to know whether the bag is likely to end up in the incinerator; about 10 percent

demand, such products are not readily available in the U.S.

Part of the problem has been the paper industry's reluctance to admit, for liability reasons, that dioxin in conventionally bleached paper might be a potential health problem. Another part of the problem is the industry's belief that consumers will buy only white paper products. Yet the industry's own poll has told them that if consumers knew about dioxin contamination, they'd avoid buying bleached paper products.

Another partial solution to dioxin contamination is the use of

of all trash does. Most plastics have a high probability of releasing noxious chemicals into the air during combustion. So, if it is going to burn, make it paper.

The ideal solution is no bag at all — at least no paper or plastic one. There are a growing number of suitable cloth bags available. Among them:

❑ **Co-op America** (c/o Order Service, 10 Farrell St., South Burlington, VT 05403; 802-658-5507, 800-456-1177) offers the "Carry-A-Ton" ($29 plus $4 shipping), a large, extra-sturdy tote with extra pockets.
❑ **Greenpeace** (P.O. Box 77048, San Francisco, CA 94107; 415-474-6767, 800-456-4029) sells a cotton shopping bag ($22 plus $4 shipping).
❑ **Save Our EcoSystems** (541 Willamette St., Suite 102, Eugene, OR 97401; 503-484-2679) sells the "Eco-Shopper," a 15" X 15" bag hand loomed in Guatemala ($16 plus $4 shipping).
❑ **Seventh Generation** (10 Farrell St., South Burlington, VT 05403; 800-456-1177) offers 100 percent cotton mesh bags, with sturdy handles, that expand and fold up small enough to fit in your pocket ($8.95 for two, $16.95 for four, plus shipping).

The next best solution is to bring back the paper or plastic bags you get at the store and reuse them on your next shopping trip. There is no reason that these bags cannot withstand dozens of shopping trips. Consider this: If every grocery bag were used twice, we would use only half the number of bags each year; use each bag five times, and we've cut bag use by 80 percent. That would all but dispose of the paper-versus-plastic-bag debate.

recycled paper. Use of gaseous chlorine in pulp mills has generally been associated with the creation of dioxins. However, the bleaching processes of many recycled paper companies are not believed to create dioxins. This is because paper recyclers do not use wood pulp and bleach with a weaker form of chlorine. Moreover, heat is a significant factor in the formation of dioxins; recycled paper is manufactured at a lower temperature than virgin paper.

Alternative bleaching techniques are being used by paper mills in Europe and by some recyclers in the U.S. Oxygen, peroxide, and sodium hydroxide are being used instead of chlorine. Although the cost of converting a paper mill from chlorine to other bleaching methods is high, there may be considerable savings in the cost of bleaching chemicals—not to mention the added value of saving human health and the environment.

There is no doubt that the answer to the problem rests with the shopping choices made by Green Consumers. Paper companies will not dare change their financially successful paper recipes unless they know that you won't continue to buy their products. Milk bottlers will continue to use dioxin-contaminated paper cartons unless they know that you will switch brands—ideally, to milk in glass bottles.

You can help deliver that message by buying the recycled and unbleached products currently available. Here are several sources:

## Unbleached Coffee Filters

❑ Permanent gold mesh coffee filters are available from kitchen supply, gourmet coffee shops, and department stores. One brand is **Swissgold**.

❑ Reusable cotton coffee filters can be used for up to one year. They are available from **Clothcrafters** (Elkhart Lake, WI 53020; 414-876-2112); **Earthen Joys** (1412 11th St., Astoria, OR 97103; 503-325-0426); and **Seventh Generation** (10 Farrell St., South Burlington, VT 05403; 802-658-5507, 800-456-1177).

❑ Unbleached paper coffee filters are available from many coffee retailers and kitchen supply stores, and directly from the manufacturers: **Ashdun Industries** (1605 John St., Fort Lee, NJ 07024; 201-944-2650), sold under the "C.A.R.E." brand; **Melitta U.S.A.** (1401 Berlin Rd., Cherry Hill, NJ 08003; 609-428-7202, 800-451-1694); and **Natural Brew** (P.O. Box 1007, Sheboygan, WI 53082; 414-459-4160).

## Unbleached Paper Towels and Toilet Paper

There are few brands of unbleached paper towels and toilet paper available directly to consumers. However, several companies manufacture them for commercial customers, including **Scott Utility Paper Towels, Crown-Zellerbach, Kimberly Clark,** and **Ft. Howard Corporation. Ashdun Industries** distributes a line of

"C.A.R.E." paper products through supermarkets. Another source is **Scott Supply, Inc.** (700 Conger, Eugene, OR 97402; 503-342-5473), which sells unbleached "natural towels."

## Unbleached Office Papers

❑ **Conservatree Paper Company** (10 Lombard St., Suite 250, San Francisco, CA 94111; 415-433-1000, 800-522-9000) offers fine writing and printing papers that are unbleached and made of 100 percent recycled materials. Conservatree, however, sells only in case-lot quantities.

❑ **Earth Care Paper Co.** (100 S. Baldwin, Madison, WI 53703; 608-256-5522) offers unbleached writing paper and envelopes, including three-ring notebook paper.

❑ **S.O.S. Recycled Paper** (Save Our EcoSystems, Inc., 541 Willamette St., Suite 315, Eugene, OR 97401; 503-484-2679) sells Earthtone paper, which is unbleached and 100 percent recycled.

# Household Cleaners

We spend hours every week cleaning our homes. Fortunately, the household products industry has taken some of the drudgery out of the work, with a long list of improvements and innovations to the old-fashioned scrub brush and pail of soapy water. But many of today's products are better suited to a laboratory than a washroom; they are filled with a long list of chemicals, some of which can be hazardous to consumers. Many other products do their damage when they are disposed of—usually down the drain.

And, of course, there is the issue of packaging. Given the increased competition for space on supermarket, drugstore, and hardware stores shelves, cleaning products manufacturers have adopted many of the same marketing techniques as food manufacturers. The inevitable result, of course, is wasteful—and expensive—overpackaging. (See "The Problem of Packaging.")

Not that you need return to the scrub brush to be a Green Consumer. There are many good products available and a growing number of alternative packaging schemes that do not require your purchasing a new container and spray (or other application) device each time you shop. The key is to understand what's contained in the products you buy, what happens to them after you use them, and how to minimize the amount of packaging necessary to clean your house—without soiling the environment.

**Down the Drain and In Your Glass.** Let's say that you are a family of four living in a freshly painted home in a good neighborhood in a nice, clean city. Your lawn is mowed, and your car is washed and waxed. The kids put away their toys. You keep things neat as a pin and never litter.

Would you believe that your tidy little family is still contributing its share of hazardous wastes down the drain and into our air and water?

Let's look at some of those cleaners you're using to keep things looking so good:

- ❑ That oven cleaner you're using most likely contains a variety of toxic and corrosive ingredients, including potassium and sodium hydroxide and ammonia.
- ❑ The furniture polish you're using probably contains toxic and flammable ingredients such as diethylene glycol and nitrobenzene.
- ❑ Your rug and upholstery cleaners contain such corrosive poisons as naphthalene, perchloroethylene, and oxalic acid.

The list goes on and on. From drain cleaners to disinfectants, furniture polishes to fungicides, most of us use these products without question. Their colorful packages and catchy advertising slogans lull us into a false sense of security. *You should be aware that the word "nontoxic" means nothing under federal law.* (See "Say What?" page 50.) It is simply an advertising slogan, like "natural." And yet we use most of these products as if they were as harmless as tap water. While there are ingredients in these everyday cleaners that you certainly wouldn't want in your drinking water, that's exactly where some of them end up.

Liquid wastes from household cleaners, washing machines, personal hygiene products, and bathtubs are supposed to end up in water treatment plants, where they are cleansed and returned to the water cycle. Studies by the Environmental Protection Agency have found septic tanks and sanitary sewage systems chock full of toxins from household wastewater. Most household septic tanks—used by 30 percent of American homes—contain over 100 traceable chemical pollutants, mostly components from cleaning fluids, degreasers, deodorants, pesticides, and other common household products. These chemicals have a tendency to migrate through soils to groundwater supplies. Approximately half of all Americans rely on groundwater, mostly untreated, as a source of drinking water.

In a study conducted in Tacoma, Washington, for example, a number of organic compounds included on the Environmental Protection Agency's priority pollutant list were found in septic tank runoff and in adjacent wells. The chemicals were deemed to have come from paint thinners, solvents, toilet bowl cleaners, and grease removers. In New York City, where residents at one apartment complex complained about the taste of their well water, tests again found that the well contained many of the same chemical compounds found in the septic tank runoff. Tests in other parts of the country have found that residential wastewaters contain numerous toxicants and are the dominant source for some chemicals entering municipal treatment systems. Residential wastes also have been found to contribute such "heavy metals" as lead, cadmium, copper, mercury, nickel, and zinc. The principal nonfood sources of these metals in residential sewage are powdered laundry detergents, fabric softeners, liquid bleach, toilet and facial tissues, and liquid dish detergents.

**Into the Air.** Water isn't the only thing affected by many of these products. Some of these chemicals escape into the air. Take aerosols, for example. Since the use of chlorofluorocarbons in aerosol sprays has been banned in the United States in 1978, we've generally considered spray cans to be safe. But aerosol products disperse the substances they spray in tiny droplets that can be inhaled deeply into your lungs and absorbed into your bloodstream. The spray paint, oven cleaner, disinfectant, furniture polishes, and other spray products you use today will become part of your family's air supply for the next week or so, especially during those times of year that your house doesn't get much fresh air.

Spray cans also pose another serious threat: when heated, they can become explosive bombs.

## *Household Hazardous-Waste Disposal*

| Chemical Products | Hazardous Ingredients | Hazardous Properties | Disposal |
|---|---|---|---|
| Abrasive cleaners or powders | trisodiumphosphate, ammonia, ethanol | corrosive, toxic, irritant | 2 |
| Ammonia-based cleaners | ammonia, ethanol | corrosive, toxic, irritant | 2 |
| Bleach cleaners | sodium or potassium hydroxide, hydrogen peroxide, sodium or calcium hypochlorite | corrosive, toxic | 2 |
| Disinfectants | diethylene or methylene glycol, sodium hypochlorite, phenols | corrosive, toxic | 2 |
| Drain cleaner | sodium or potassium hydroxide, sodium hypochlorite, hydrochloric acid, petroleum distillates | corrosive, toxic | 3 |
| Floor and furniture polish | diethylene glycol, petroleum distillates, nitrobenzene | flammable, toxic | 3 |
| Household batteries | mercury, zinc, silver, lithium, cadmium | toxic | 1, 3 |
| Mothballs | naphthalenes, paradichlorobenzene | toxic | 2, 4 |

What can you do about these household cleaning problems? There are two basic solutions:

- ❑ Use nonpolluting alternative cleaning methods. See page 120 for nonpolluting brands and page 123 for recipes for effective homemade cleaners made from common household products.
- ❑ Dispose of these hazardous products in ways so they will cause minimum harm.

| Chemical Products | Hazardous Ingredients | Hazardous Properties | Disposal |
|---|---|---|---|
| Mothballs | naphthalenes, paradichlorobenzene | toxic | 2, 4 |
| Oven cleaners | potassium or sodium hydroxide, ammonia | corrosive, toxic | 2 |
| Photographic chemicals | silver, acetic acid, hydroquinone, sodium sulfite | corrosive, toxic, irritant | 3 |
| Pool chemicals | muriatic acid, sodium hypochlorite, algicide | corrosive, toxic | 3 |
| Rug and upholstery cleaners | napthalene, perchloroethylene, oxalic acid, diethylene glycol | corrosive, toxic | 3 |
| Toilet cleaners | muriatic (hydrochloric) or oxalic acid, paradichlorobenzene, calcium hypochlorite | corrosive, toxic, irritant | 2, 3 |

**Disposal Notes**

1. Recyclable. Take to a service station, reclamation center, or household hazardous-waste collection center. Partially used products such as paints may be exchanged.
2. Fully use these products so that no waste remains except residuals attached to the container. Containers should be rinsed with water. The container may then be disposed of at the municipal landfill; the rinse water may be reused or poured down the drain and flushed with great quantities of water.
3. These wastes should be safely stored until a hazardous-waste program is organized in your community. Fully spent currently available pesticide containers may, however, be triple rinsed and the rinse water reused according to instructions on the label.
4. Waste can be disposed of at some waste-water treatment plants where bacteria can detoxify the chemical. (Call your local treatment plant.) Do not pour on the ground.

Courtesy Environmental Hazards Management Institute, 10 Newmarket Rd., P.O Box 932, Durham, NH 03824; 603-868-1496

**Detergents and Phosphates.** Concern about phosphate in laundry and dishwashing detergents first arose in the early 1960s, when foaming and pollutant-choked streams, rivers, and lakes became one of North America's first environmental crises. Since then, eleven states have banned phosphate from detergents sold within their borders, but it is still permitted in most of the country, and still contributes to water pollution. Phosphate is added to detergents to soften water and improve detergent performance. According to the Soap and Detergent Association, detergents with phosphate yield better performance per dollar spent.

But it also yields problems for fish and other things that live in water. When phosphate, a plant nutrient, enters a body of water, algae feed on it. Too much phosphate causes algae to grow quickly. When the algae die, they cloud the water, blocking sunlight. As they decompose, they use up oxygen. The combined process, known as accelerated atrophication (to which phosphate is only one contributor) speeds up the natural aging process of these waterways and can be lethal to most aquatic life. Thus, a vital waterway can become a swamp. Phosphate was one of the contributors that resulted in the virtual death of Lake Erie in the 1970s. Phosphate bans have helped to clean up Lake Erie, as well as the Chesapeake Bay and other threatened bodies of water. According to Clean Water Action, a Baltimore-based group, the phosphate ban has saved Maryland's sewage treatment plants more than $4.4 million a year in chemicals and cartage for phosphate treatment.

When phosphate-free detergents were introduced in the 1970s, they were not embraced by consumers, due to their poor performance. Consumers used twice as much detergent or washed loads twice, then switched back to brands containing phosphate. Today's phosphate-free detergents have been vastly improved. When *Consumer Reports* tested eight phosphate and phosphate-free brands in 1987, it reported that there was negligible difference between the two. (See "Green Cleaners," on page 120, for a list of companies that sell alternative, nonpolluting cleaning products.)

**Where Phosphate Is Banned.** In some jurisdictions, phosphate is prohibited only in selected counties.

Chicago, Illinois
Connecticut (selected counties)
District of Columbia
Indiana
Maryland
Michigan
Minnesota
Montana (selected counties)
New Jersey (selected counties)
New York State
North Carolina
North Dakota (selected counties)
Ohio (selected counties)
Pennsylvania (selected counties)
South Dakota (selected counties)
Vermont
Virginia
Wisconsin

## Using Cleaning Products Safely

Here are some suggestions for using and storing household cleaning products safely, excerpted from *Guide to Hazardous Products Around the Home, 2nd Edition,* available for $8 postpaid from the **Household Hazardous Waste Project** (901 S. National, Box 108, Springfield, MO 65804; 417-836-5777). (See also the table, "Household Hazardous-Waste Disposal," on pages 116-117.)

❑ Read all labels carefully before using hazardous products. Be aware of their uses and dangers.
❑ Leave products in their original containers with the label that clearly identifies the contents. Never put hazardous products in food or beverage containers.
❑ Do not mix products unless instructed to do so by label directions. This can cause explosive or poisonous chemical reactions. Even different brands of the same product may contain incompatible ingredients.
❑ Use only what is needed for a job. Twice as much doesn't mean twice the results. Follow label directions.
❑ If you are pregnant, avoid toxic chemical exposure as much as possible. Many toxic products have not been tested for their effects on unborn children.
❑ Avoid wearing soft contact lenses when working with solvents and pesticides. They can absorb vapors from the air and hold the chemicals near your eyes.
❑ Use products in well-ventilated areas to avoid inhaling fumes. Work outdoors whenever possible. When working indoors, open windows and use an exhaust fan, making sure air is exiting outside rather than being circulated indoors.
❑ Do not eat, drink, or smoke while using hazardous products. Traces of hazardous chemicals can be carried from hand to mouth. Smoking can start a fire if the product is flammable.
❑ Clean up after using hazardous products. Carefully seal products. Properly refasten all childproof caps.

### Storing Products

❑ Make sure containers are kept dry to prevent corrosion. If product container begins to corrode, place it in a plastic bucket

### *Before You Buy*

Here are some things to consider before you purchase household cleaning supplies:

❑ Do I really need this product?
❑ Does it contain ingredients of concern?
❑ Is there a safer alternative?
❑ Will I be able to dispose of it properly?
❑ Can I safely store this product in my home?

with a lid and clearly label the outside container with contents and appropriate warnings.
❑ Store volatile chemicals or products that warn of vapors or fumes in a well-ventilated area, out of reach of children and pets.
❑ Store rags used with flammable products (including furniture stripper, paint remover, and gasoline) in a sealed container.
❑ Keep products away from heat, sparks, flames, or sources of ignition. This is especially important with flammable products.
❑ Store gasoline only in approved containers, away from all sources of heat, flame, or sparks, in a well-ventilated area.
❑ Know where flammable materials are located in your home and know how to extinguish them. Keep a working fire extinguisher in your home. (For more on buying fire extinguishers, see page 210.)

## GREEN CLEANERS

A growing number of nontoxic, biodegradable, and often cruelty-free cleaning products are available, some sold in health food stores; a few may even be found in general supermarkets. (In addition, consider the alternative cleaning techniques beginning on page 123 that use many of the natural ingredients already in your pantry.) One nationally available brand is **Ecover** cleaning products, which are made of organic ingredients that degrade within five days; they are phosphate free, not petroleum based, and are not tested on animals. Ecover products include laundry detergent, fabric conditioner, wool wash liquid, dishwashing liquid, floor soap, toilet cleaner, and bath and kitchen cleaner. They are available from several of the sources below.

Don't overlook a number of time-tested cleaning products that are readily available on supermarket shelves. Such mild soaps as **Ivory Snow, Fels Naptha, Arm & Hammer**, and **Bon Ami** work well for a variety of cleaning needs.

❑ **AFM Enterprises Inc.** (1140 Stacy Court, Riverside, CA 92507; 714-781-6860) sells biodegradable cleaning products, including two all-purpose cleaners, carpet shampoo, and a mildew-control formulation.

❑ **CHIP Distribution** (P.O. Box 704, Manhattan Beach, CA 90266; 213-545-5928) sells CON-LEI cleaners, laboratory tested and proved to be nonflammable, nonexplosive, noncaustic, and biodegradable within seven days. The cleaners are available as concentrates, which, when diluted, can be used for cleaning everything from pots and pans to marine bilges and heavy equipment. The products are available through CHIP Distribution or by mail through **Co-Op America**, 2100 M St. NW, Suite 310, Washington, DC 20036; 202-872-5307.

❑ **Cloverdale** (P.O. Box 268, West Cornwall, CT 06796; 203-672-0216) distributes "Cloverdale," a biodegradable cleaner and degreaser, good for everything from washing clothes to cleaning the car. Cloverdale is sold in concentrated form and can be diluted with different amounts of water for various cleaning tasks. It is available in quarts, gallons, and 5-gallon and 55-gallon drums through **The Sprout House**, 40 Railroad St., Great Barrington, MA 01230; 413-528-5200.

❑ **The Compassionate Consumer** (P.O. Box 27, Jericho, NY 11753; 718-445-4134) offers Ecover floor soap and toilet cleaner, Golden Lotus all-purpose cleaner, liquid dishwashing soap, laundry detergent, fabric softener, and bleach.

❑ **Livos Plantchemistry** (641 Agua Fria St., Santa Fe, NM 87501; 505-988-9111) produces natural ingredient products, including "Avi—Soap Concentrate" and "Latis—Natural Soap" for all-purpose cleaning needs, as well as "Lavenos—Hand Soap," "Snado—Leather Polish," "Gleivo—Liquid Furniture Wax," and "Bilo—Floor Wax." A catalog is available.

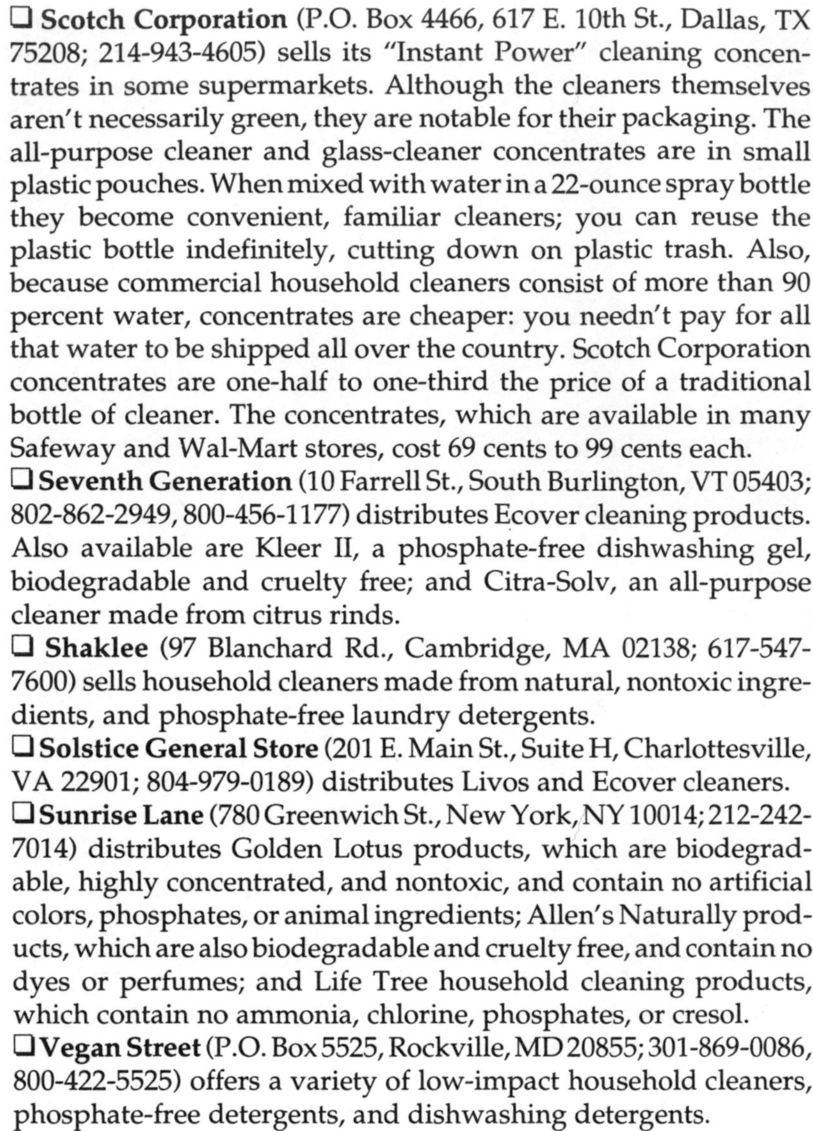

❑ **Scotch Corporation** (P.O. Box 4466, 617 E. 10th St., Dallas, TX 75208; 214-943-4605) sells its "Instant Power" cleaning concentrates in some supermarkets. Although the cleaners themselves aren't necessarily green, they are notable for their packaging. The all-purpose cleaner and glass-cleaner concentrates are in small plastic pouches. When mixed with water in a 22-ounce spray bottle they become convenient, familiar cleaners; you can reuse the plastic bottle indefinitely, cutting down on plastic trash. Also, because commercial household cleaners consist of more than 90 percent water, concentrates are cheaper: you needn't pay for all that water to be shipped all over the country. Scotch Corporation concentrates are one-half to one-third the price of a traditional bottle of cleaner. The concentrates, which are available in many Safeway and Wal-Mart stores, cost 69 cents to 99 cents each.

❑ **Seventh Generation** (10 Farrell St., South Burlington, VT 05403; 802-862-2949, 800-456-1177) distributes Ecover cleaning products. Also available are Kleer II, a phosphate-free dishwashing gel, biodegradable and cruelty free; and Citra-Solv, an all-purpose cleaner made from citrus rinds.

❑ **Shaklee** (97 Blanchard Rd., Cambridge, MA 02138; 617-547-7600) sells household cleaners made from natural, nontoxic ingredients, and phosphate-free laundry detergents.

❑ **Solstice General Store** (201 E. Main St., Suite H, Charlottesville, VA 22901; 804-979-0189) distributes Livos and Ecover cleaners.

❑ **Sunrise Lane** (780 Greenwich St., New York, NY 10014; 212-242-7014) distributes Golden Lotus products, which are biodegradable, highly concentrated, and nontoxic, and contain no artificial colors, phosphates, or animal ingredients; Allen's Naturally products, which are also biodegradable and cruelty free, and contain no dyes or perfumes; and Life Tree household cleaning products, which contain no ammonia, chlorine, phosphates, or cresol.

❑ **Vegan Street** (P.O. Box 5525, Rockville, MD 20855; 301-869-0086, 800-422-5525) offers a variety of low-impact household cleaners, phosphate-free detergents, and dishwashing detergents.

## Natural Cleaners

The greenest cleaners of all reside in many of the products you probably have in your kitchen: baking soda, vinegar, lemon juice, vegetable oil, borax, and good old hot water, for example. While not necessarily as alluring as a "magic-bullet" spray product that promises to clean everything under the sun in mere seconds, most of the alternative cleaning methods below are just as effective, far less expensive, and nonpolluting.

**Aerosols**: Even though they no longer contain CFCs, the nonrecyclable containers are a waste of resources and pollute the environment; also, the microscopic aerosol-propelled particles may be harmful to the lungs, heart, and central nervous system when inhaled. Avoid aerosols; if you must spray use pump products.

**Air fresheners**: These mask smells by coating nasal passages and deadening nerves to diminish the sense of smell. Don't use commercial air fresheners. Instead:

- ❑ find source of odors and eliminate them
- ❑ keep house and closets clean and well-ventilated
- ❑ set out 2-4 tablespoons of vinegar or baking soda in open dishes
- ❑ use house plants; they are good air purifiers
- ❑ boil herbs and spices for natural fragrance

**All-purpose cleaners**: Ammonia and chlorine are in many all-purpose cleaners. These form deadly chloramine gas when mixed. Ammonia itself can be harmful to lungs, while chlorine can form cancer-causing compounds when released into the environment. Make your own cleaner by mixing two teaspoons of borax and one teaspoon of soap in one quart of water in a rinsed-out spray bottle. Or use a half cup of washing soda (hydrated sodium carbonate) in a bucket of water; it works on all but aluminum surfaces.

**Carpet deodorizers**: Sprinkle baking soda or cornstarch on carpet, using approximately one cup per medium-size room. Vacuum after 30 minutes. Or mix two parts cornmeal with one part borax, sprinkle liberally, leave one hour, and vacuum.

**Dishwashing liquids**: Most dishwashing liquids are detergents, derived from petroleum; they are nonbiodegradable and usually contain chemical additives such as artificial fragrances and colors. (Detergents also cause more child poisonings than any other household product). Use liquid or powdered soap such as Ivory (add two to three teaspoons of vinegar for heavy soil). In dishwashers, use equal parts borax and washing soda (hydrated sodium carbonate); increase the proportion of soda for hard water.

**Disinfectants**: Most disinfectants are a mix of toxic chemicals including phenol, formaldehyde, cresol, ammonia, and chlorine. Instead, mix one-half cup borax in one gallon hot water. This was tested in a California hospital for one year and met all state germicidal requirements, according to the Clean Water Fund.

**Drain cleaners**: The lye, hydrochloric acid, and sulfuric acids in drain cleaners can burn human tissue, causing permanent damage. If not used according to instructions, they can explode; these are especially dangerous around children. Prevention is the best strategy here. Never pour grease down a drain, always use a drain sieve or hair trap, and clean the metal screen or stopper mechanism regularly. Once a week, as routine maintenance, plug the overflow drain with a wet rag, pour a quarter cup baking soda down the drain, follow with one-half cup vinegar, and close the drain tightly until fizzing stops. Flush with one gallon boiling water. For persistent clogs, use a rubber plunger or a metal drain snake, available at hardware stores.

**Flea and tick control**: Most pesticides used for flea and tick control have never been adequately tested for safety. These products rub off the pet onto people and furniture, exposing your family to the risk of cancer and other diseases. Feed two tablespoons of brewer's yeast and a clove of raw garlic to pets daily. Also, pet dips and sprays containing de-limonine gas derived from citrus extracts safely repel pests. Pyrethrin powders made from ground chrysanthemums sprinkled on the carpet, then vacuumed, prevent further infestation. Insecticidal soaps are biodegradable and nontoxic, and kill fleas, ticks, and lice instantly. Wash pets with warm soapy water, dry thoroughly, and use the following rinse: one-half cup fresh or dried rosemary in one quart of boiling water; steep 20 minutes, strain, and cool. Spray or sponge onto pet and allow to dry (don't towel dry). Also, organic repellents made from distillates of cedarwood, orange, eucalyptus, and bay are available for house sprays and flea collars. (See "Pets and Pet Supplies.")

**Floor cleaners**: Dull, greasy film on no-wax linoleum can be washed away with one-half cup white vinegar mixed into one-half gallon warm water.

**Floor and furniture polish**: Many wood polishes contain phenol, which causes cancer in laboratory animals. Ingesting one thimbleful of phenol can cause symptoms ranging from circulatory collapse to death. Residual vapors contaminate the home long after use. Also, wood polish may cause severe skin irritation. Mix a one-to-one ratio of vegetable oil and lemon juice or vinegar into a solution and apply a thin coat. Rub in well. On unwaxed wood, use vegetable oil and lemon oil to replenish shine.

**Glass cleaner**: Commercial products emit ammonia mist, which enters the lungs. Ammonia is a poison—use it only when other cleansers won't do the job. *Never mix ammonia with bleach or commercial cleansers—deadly fumes may result*. First, use alcohol to clean the residues left from commercial glass cleaners. Then clean the glass with a mixture of half white vinegar and half water.

**Insect repellents**: Commercial insect repellents contain a variety of toxic chemicals. There are several safe and effective alternatives, such as burning citronella candles. Or plant sweet basil around the patio and house to repel mosquitoes. Still another method is to blend six cloves of crushed garlic, one minced onion, and one tablespoon soap in a gallon of hot water. Let sit one or two days, strain, and apply with a spray bottle.

**Laundry products**: Most laundry powders are nonbiodegradable detergents. Use phosphate-free, biodegradable detergents, or better yet, switch to soap flakes (such as Ivory). When switching, wash items once with washing soda (hydrated sodium carbonate) only. This eliminates detergent residues that might react with soap to yellow fabrics. Boost soap products with washing soda; this will brighten all washable fabrics and costs less than bleach.

**Metal polishes**: The fumes from phosphoric and sulfuric acids and ammonia contained in commercial metal polishes contribute unnecessarily to indoor air pollution. For silver, soak 10 to 15 minutes in one quart warm water, one teaspoon baking soda, one teaspoon salt, and a small piece of aluminum foil; wipe with a soft cloth. Or rub with a paste of baking soda and water. For aluminum, dip cloth in lemon juice and rinse with warm water; or soak overnight in a mixture of vinegar and water, then rub. For brass, mix equal parts of salt and flour with a little vinegar, then rub. For chrome, rub with undiluted vinegar. For copper, rub with a paste of lemon juice or vinegar, salt, and flour, or hot vinegar and salt. For gold, wash in lukewarm soapy water, dry, and polish with a chamois cloth.

**Mold and mildew cleaners**: These may contain pesticides. Instead, make a concentrated solution of borax or vinegar and water, and clean affected areas. For mildew, try a mixture of lemon juice or white vinegar and salt. Borax will inhibit mold growth.

**Mothballs**: These often contain p-dichlorobenzene, a known carcinogen. Use cedar chips or herbal sachets; store woolens in cedar-lined closet or trunk. Moth eggs can be destroyed by running the item through a hot dryer. (Be careful, however: a hot dryer can damage or shrink some clothes. If you are uncertain, read the washing instructions on the clothing label.)

**Oven cleaners**: The basic ingredient in oven cleaners is lye, a powerful caustic that can burn and disfigure. Exposure can scar your lungs or cause blindness if splashed on eyes. Prevent the need for oven cleaning by avoiding overfilling pans, and scrape up spills as soon as food is cool enough to handle. When cleaning, remove remnants of charred spills with a nonmetallic bristle brush. Clean oven with a paste of baking soda, salt, and hot water. Or sprinkle with dry baking soda and scrub with a damp cloth after 5 minutes. (Don't let baking soda touch wires or heating elements.)

**Pesticides**: These contain some of the most toxic chemicals around. Many have been linked to birth defects, leukemia, and cancer. As a general rule, keep kitchen, floors, and garbage pails clean to eliminate pests' food supplies and remove clutter to eliminate nesting areas. For ants, sprinkle cream of tartar, red chili powder, dried peppermint, or boric acid where they enter. For cockroaches and silverfish, use equal parts baking soda and powdered sugar. (The sugar attracts them, the baking soda kills them.) For fleas, use flea combs or herbal flea powders on pet and keep house thoroughly vacuumed. For slugs and snails, place copper-sheeting barriers around sensitive plants. For houseflies, use sticky untreated flypaper; or make your own with honey and yellow paper. For mice and rats, use mousetraps or mix one part plaster of Paris with one part flour and some sugar and cocoa powder; sprinkle where rodents (but not children) will find it. As for spiders, leave them alone—they eat other insect pests.

**Toilet cleaners**: These contain chlorine and hydrochloric acid, which can burn skin and eyes. Instead, use soap and borax; remove stubborn rings and lime buildup with white vinegar. Another cleaner is baking soda sprinkled into the bowl; drizzle with vinegar and scour with a toilet brush.

Four excellent resources for further information on alternative cleaners are:

- ❑ "Home Safe Home," a chart of household alternatives with a disposal checklist. Available for $1 (25 for $10) from the **Clean Water Fund**, New Jersey Environmental Federation, 808 Belmar Plaza, Belmar, NJ 07719; 201-280-8988.
- ❑ *Stepping Lightly on the Earth: Everyone's Guide to Toxics in the Home*, a thorough and informative guide to problems with toxics and alternatives. Available free from **Greenpeace**, 1436 U St. NW, Washington, DC 20009; 202-462-1177.
- ❑ *The Nontoxic Home*, by Debra Lynn Dadd (Jeremy Tarcher, 5858 Wilshire Blvd., Los Angeles, CA 90036; 213-935-9980. 1986, $9.95), provides background information on how to decide which home products are healthy and safe.
- ❑ *Nontoxic and Natural and Earthwise*, by Debra Lynn Dadd (Jeremy Tarcher, 5858 Wilshire Blvd., Los Angeles, CA 90036; 213-935-9980. 1990, $9.95), rates the toxicity of 1,200 brand-name products from cosmetics and foods to office supplies and building materials. Also included are lists of mail-order sources for safe products, and do-it-yourself formulas for common household products.

## Water Safe to Drink

There have been enough unsettling stories about unsafe drinking water to make one hesitate before having a refreshing glass of *agua pura*. Indeed, with the dozens of exotic and noxious chemicals—from household waste, industrial dumping, and agricultural runoff—seeping into our nation's groundwater supplies, the potential for danger is high, and getting higher.

It's no wonder, then, that a growing number of people have turned to bottled water, or have used some water-filtering gadgetry in an attempt to thwart such dangers. And many of these resources do produce really good-tasting water.

But that doesn't necessarily mean that bottled waters are better for you than tap water—or that they are better for the environment.

**The Myths of Bottled Water.** Once the beverage of a small but dedicated corps of purists, bottled water is now the beverage of choice for millions of Americans. According to the International Bottled Water Association, Americans drank more than 1.7 billion gallons of bottled water in 1989, at an average cost of $1.15 per gallon (compared with an average price of $1.28 per *thousand* gallons of tap water, making bottled water nearly 900 times more expensive). And their reasons for drinking bottled water—whether the generic kind that comes in a plastic jug or the more expensive name-brand varieties—contain equal portions of fact, fantasy, and fashion. But to understand them requires an explanation of the three basic types of water on the market:

❑ **Processed or purified water:** By removing minerals, sterilizing the water through a process called ozonation, and replacing the minerals, bottlers are capable of creating waters that rival Mother Nature's finest in purity. Processed water, often made simply from city tap water, has long constituted the largest share of the bottled water market. Giant national companies like **Foremost-McKesson** and **Bordon's** have the lion's share of this market.

❑ **Natural still water:** Also known as "mineral water," this category includes many of the popular domestic and imported brands. Still waters are commonly distributed in five-gallon bottles suitable for use with office coolers or with a simple plastic hand pump for the home. Beware, however: some waters sold in five-gallon containers are purified, not natural waters. Natural still waters are extracted from underground springs; some are bottled at the source, while others are transported by truck or rail for bottling. Popular brands are **Evian, Mountain Valley,** and **Deer Park,** as well as a few hundred lesser-known regional brands.

❑ **Natural sparkling water:** These are among the priciest bottled waters and include such popular brands as **Perrier, Apollinaris,** and **Poland Spring Sparkling Water**. What makes these waters sparkle is carbon dioxide.

The distinction between "natural" and "purified" waters is at times rather murky. Even "natural" waters are sometimes "purified" through ozonation, which involves injecting heavy oxygen molecules into the water to kill bacteria. Some sparkling waters are actually natural still waters with carbonation added, and there are squabbles within the industry whether this procedure should deprive the water of its designation as "natural." Poland Spring water is extracted from a source in the White Mountains in Maine, but its carbonation is trucked in from a well in Colorado.

All of this is made more confusing by the fact that there are few laws to require that bottled water be any purer than the stuff that comes out of your kitchen faucet. While the bottled water industry may claim that its product "is a most highly regulated and monitored drinking water supply," the regulations and monitoring still don't deal with some serious problems. The regulations enforced by the federal government set standards for bottled water that are exactly the same as for tap water—and those tap water standards are not exceptional. The rules set "tolerance" levels for some very potent chemicals, but they don't specify any limits for a wide variety of synthetic compounds called "organics"—an alphabet soup of carcinogens (which cause cancer) and mutagens (which cause birth defects) like PBB, PVC, THM, and many pesticides. About 700 organics have been measured in drinking water so far; many of them have never been tested for toxicity. True, bottled water must come from protected sources (although that can include a local faucet), be bottled in facilities regulated as food plants, be processed using manufacturing practices approved by the federal government, and provide adequate labeling. But bottlers can abide by all these rules and still not even test their water for organics.

The purification and distillation processes used in making bottled waters do kill bacteria. But during the distillation process, when water is vaporized so that minerals can be removed, some of the deadly organics remain in the water vapor and get reconstituted right back along with everything else. In areas where polluted tap water is the basis of bottled water, there's an excellent chance that the bottled version will be just as polluted as the original.

Even waters from faraway mountain streams don't escape pollution. "Nonpoint" sources of pollution—where rainwater collects pollutants and sends them into streams and rivers—put chemicals into the water cycle, and they never seem to leave. Small towns that have no industry and are miles from cities have found distressing levels of pollutants in their water supplies. Product claims about product purity and cleanliness may hold much credibility in the 1990s, when even the rain is dirty. And while many bottlers are quick to push impressive chemical analyses at you, you will be hardpressed to find one in which the bottler has analyzed the water's level of organics.

**Is Bottled Water "Green'?** What about the environmental impact of bottled water? Unfortunately, from that perspective bottled waters don't look good. For one thing, the resources used in packaging, and the energy costs used in shipping water around the country and overseas, are tremendous, given that most of us have adequate resources at hand. While there are no available figures on the energy costs (and associated pollution) resulting from this activity, you can be sure that it isn't insignificant.

And then there's the trash. Plastic and glass beverage bottles account for the second largest source of solid waste, and bottled water packaging contributes to that pile of nonbiodegradable trash.

So, bottled water is not a particularly green product, no matter how many pollutants you feel you may be avoiding by drinking it.

Still, if you *know* your water is unhealthy, bottled water may at least give you peace of mind. If you must drink it, buy water that is from sources nearby, minimizing transportation energy costs, and is packaged in recyclable containers such as glass. Your best bet may be to rent a dispenser that holds recyclable five-gallon glass jugs of local water. There are many home-delivery services that will bring refillable containers of water to your house on a weekly basis; check the Yellow Pages for listings.

**Home Purification Products.** Ever since we discovered clean water in a bottle, we've been trying to find ways of getting around paying a dollar or so per quart for it. But the alternative may cause more problems than it solves.

Most home water filters are not very effective. In fact, some have been found to *add* pollutants to water. The problem is caused by the activated carbon used in the filtering process of many systems. Bacteria stick to the carbon—as they are supposed to—but they also grow and multiply, especially overnight and other times when the tap isn't being used for a long period. Turning on the water releases the bacteria into the sink—or into your glass. To counter the problem, some manufacturers add metallic silver to their filters, which effectively retards bacterial growth. Unfortunately, it also adds a significant new contaminant to the water: high levels of silver can pose their own health problems.

Besides activated carbon filters, home water treatment devices use other methods to cleanse water: reverse osmosis, ion exchanging, and distillation, among others. Prices range from about $100 to over $1,000. It is important to keep in mind that all of these systems must be well maintained to work properly. Failure to do this may result in additional pollutants being added to the water.

The bottom line is that most filters don't screen out the really dangerous pollutants for which they were intended, most important, the organics, which are being found increasingly in water systems around the country. There are no federal regulations governing home water treatment devices, leaving it up to you to determine whether the water that comes out of the filtering device is actually better than the water that goes in.

If you are thinking about a home water treatment system, consider the following suggestions:

*[continued on page 134]*

## *What's Wrong with Your Water?*

| Sensory Clue | Symptom | Probable Cause | Contaminant | Treatment Options |
|---|---|---|---|---|
| Sight | staining and pitting of teeth | naturally occurring or excess addition of fluoride at treatment plant | fluoride | distillation, ion exchange, activatec carbon, activated alumnine |
| Sight | turbidity (cloudiness) | organic or susp ended matter; defective well screen or inade-quate public treatment | dirt, sand, clay, silt, organic matter | new well screen, carbon or sand trap chlorination |
| Sight | blue-green stains on sink and porcelain fixtures | corrosive water reacts with brass or copper pipes or fittings | brass, copper | pH adjustment by calcite or soda ash and filtration |
| Sight | brown-red water, stains and discolored clothing | naturally occurring iron, iron bacteria, leached from old pipe | iron | manganese green sand filter, chlorination filtration |
| Sight | yellow water | water passing through peaty soil and vegetation | tannins (humic acid), yellow iron | anion exchange, chlorination |
| Sight | black staining of fixtures and laundry | naturally occurring, usually found with iron | manganese | manganese green sand filter |
| Sight | blackening and pitting of stainless steel sink | excess salt concentration, road salting, improper brine backwash from water soft-ener treatment | chloride | R-osmosis, ion exchange |
| Sight | milky water | excessive air or particles in water | particles, air | service filters, let water stand, clean hot water heater |
| Smell | fishy, sweet, perfume odor | industrial chemicals or waste | s, v* | locate and remove source, activated carbon with monitoring |
| Smell | oil or gas smell | gasoline or oil spill, leak (e.g. , under ground gas station tank) | g, s* | locate and remove source, activated carbon with monitoring |

| Sensory Clue | Symptom | Probable Cause | Contaminant | Treatment Options |
|---|---|---|---|---|
| Smell | musty, earthy smell | algae, vegetation | organic matter (leaves, algae) | activated carbon, remove and correct problem, seek different source |
| Smell | chlorine smell | excess chlorination or inadequate dechlorination | chlorine | activated carbon, contact local treatment department |
| Smell | detergent smell (foamy water) | waste water discharge into water supply | foaming agents, dilute sewage | eliminate source, chlorination |
| Smell | rotten egg smell | sewage, sulfate reducing bacteria | hydrogen sulfide (gas) | chlorination, manganese green sand filter |
| Taste | alkali taste | high mineral content in private well and/or any water supply | minerals | R-osmosis, ion exchange, distillation |
| Taste | metallic taste | corrosive water causing leaching, high mineral content (naturally occurring) | iron, manganese | calcite, water softener, chlorination, aeration |
| Taste | salty, brackish taste | road salting, salt water intrusion, naturally occurring in arid regions, water softener backwash | sodium chloride, sulfate, inorganic salts | distillation, R-osmosis, ion exchange |
| Taste | sharp chemical taste or odor, or "oily" consistency | industrial waste activities, waste disposal | h, p, s* | locate and remove source, activated carbon with monitoring |

* **Contaminant Codes** (Common contaminants not specified on the chart are listed behow by groups)
**M**: Minerals: arsenic, asbestos, cadmium, chromium, cyanide, lead, mercury
**G**: Gasoline: benzene, toluene, p-dichlorobenzene, chlorobenzene
**V**: Volatile organics: trichloroethylene; carbon tetrachloride; vinyl chloride; 1,2-dichloroethane; benzene; p-dichlorobenzene; 1,1-dichloroethylene; 1,1,1-trichloroethylene; chloroform.
**S**: Semi-volatiles: cresolic and phenolic compounds, naphthalenes, phthalates.
**P**: Pesticides: Aldrin, Toxaphene, Chlordane, Heptachlor, Dieldrin, Endrin, Lindane, Methoxychlor, EDB, Alachlor.
**H**: Herbicides: 2,4-D; 2,4,5-TP (Silvex); Dinoseb.

Courtesy Environmental Hazards Management Institute, 10 Newmarket Rd., P.O Box 932, Durham, NH 03824; 603-868-1496

❑ **Get your water tested by a local service.** They are usually listed in the Yellow Pages under "Environmental Services." This will help you decide whether you even need a system in the first place. You may be surprised to find that your water is in pretty good shape.

❑ **Read up on the system you are considering.** Make sure you understand what pollutants it removes, how often you must change the filter, and whether you can do it yourself.

❑ **Ask for a no-obligation 30-day tryout.** Many companies will let you do this so that you can see for yourself whether the cost of the device is worth it.

(See also page 205 for information about home water-saving devices.)

# G·A·R·D·E·N AND P·E·T S·U·P·P·L·I·E·S

There is nothing more "green" than growing a garden. Whether it is a window box hanging from an apartment window or an acre behind your house, gardens provide a host of benefits to your psyche and your refrigerator. At least 69 million Americans did some type of gardening and lawn care during 1988, according to the National Gardening Association, and the numbers are growing each year. According to a 1987 Louis Harris poll, gardening is the number-one recreational activity in the United States. And all that gardening uses a lot of chemicals: a 1980 study by the National Academy of Sciences showed that residential lawns and gardens received as much as 10 pounds of chemicals per acre—compared with about 2 pounds per acre for soybean crops.

Unlike many of our recreational activities, this one is good for the environment. All growing vegetation absorbs carbon dioxide, which otherwise is released into the atmosphere and is a major contributor to greenhouse warming (see "Seven Environmental Problems You Can Do Something About"). The more vegetation that exists, the less carbon dioxide is released into the atmosphere. And, of course, if you grow vegetables, the more you grow the less you'll have to buy, alleviating your contributing to energy costs and pollution associated with most modern agriculture.

But gardening also can be far from green, depending on how it is done. Although gardening supply and chemical companies have introduced a lengthy list of products to make gardening more productive and less work, many of these products are not good for you, your children, your pets, and other living things. Even if you aren't growing anything edible—by humans, at least—you may still be harming other creatures, some of whose presence could be protecting your garden from a variety of less-than-welcome pests. The fact is, there is a variety of good-quality products and techniques you can use to have both a green thumb and a green environment.

## Problems with Pesticides

Most gardeners face a dilemma: Do you deploy an arsenal of chemicals on your plants, or do you go "chemical free" and hope for the best? In the past, most people opted for the chemicals. And there have been many to choose from—an estimated 83 brands of pesticides offering between 300 and 400 individual products are sold on retail shelves of gardening, hardware, and grocery stores, as well as by mail order—for which we spend nearly $1 billion a year, according to the National Gardening Association. None of which includes the billions of pounds of pesticides dumped on crops by American farmers (see "How Safe Is Our Food?"). According to the Environmental Protection Agency, suburban lawns and gardens probably receive the heaviest applications of pesticides of any land in the United States.

The fact is, the safety of a substantial number of these products has yet to be established or already has been seriously questioned. A 1972 federal law gave to the EPA the task of re-registering all pesticides then on the market. The re-registration process includes a detailed examination of data on safety as well as both short-term (acute) and long-term (chronic) health effects. But due to weak enforcement and bureaucratic entanglement, only about 120 of the 600 principal ingredients in commercially available pesticides have been registered so far. So, don't assume that because a product is available in your local hardware or garden store, it has undergone rigorous scrutiny by the government. There's an 80 percent chance that it hasn't. Moreover, some pesticides that were once widely used have now been banned or severely restricted, including DDT, chlordane, aldrin, heptachlor, dieldrin, lindane, silvex, and 2,4,5-T. If you are storing pesticides that you haven't used in two or more years, there's a chance they might contain one or more of these ingredients. (See box, "Chemicals to Watch Out For.")

What are the effects of all these chemicals on you and on the environment? That the effects can be deadly has been known for years—since 1962, when Rachel Carson published her landmark book *Silent Spring*, which exposed to the general public for the first time the horrid effects of pesticides on water, soil, and air, and the wildlife and people they support. As Carson put it: "For the first time in the history of the world, every human being is now subjected to contact with dangerous chemicals, from the moment of conception until death." But since *Silent Spring*, things haven't necessarily improved; in many ways they've gotten worse. There are greater amounts of a greater number of chemicals being used than ever before.

The problems aren't limited to a few well-publicized toxic waste dumps. The pesticides you use at home—and which are used in parks, golf courses, and other nonagricultural settings—are carried by the wind and are transported through the soil and sewer systems into rivers, lakes, and streams.

And that's only the beginning. Birds, squirrels, rabbits, dogs, cats, insects, fish, and many other creatures dine on many of the things contaminated by these chemicals—an even the little patch of green in your backyard can be contributing to the problem.

Unfortunately, many of these chemicals cause health problems for humans, too, although there is considerable disagreement among government scientists about the cancer-causing potential of many chemical ingredients; some pesticides are rated as "probable" or "possible" carcinogens. Only three ingredients on the chart on pages 150-151 have been evaluated for potential carcinogenicity by the Environmental Protection Agency: **rotenone** (no consensus reached, but studies did produce tumors in rats), **propoxur** (rated a "probable" carcinogen), and **benomyl** (a "possible" carcinogen). Some environmentalists and health officials take a stronger position on the cancer-causing properties of some other ingredients, including **benomyl, dicofil, glyphosate, malthion,** and **2,4-D**.

All of which leaves much of the guess work about the safety of pesticides up to you.

## *Chemicals to Watch Out For*

Since 1977, the federal government has determined which products it considers too risky for general consumer use. These products may legally be applied by certified applicators licensed by the state. These products' labels contain the warning RESTRICTED USE PESTICIDE. If you have any pesticide with this label, you should consider it a hazardous waste. Seal up the package carefully and deliver it to a hazardous-waste collection site.

Many basements, garages, and garden sheds contain products purchased decades ago. A package of pesticide purchased before 1983 won't necessarily be labeled RESTRICTED USE, even if it is now considered too risky for general consumer use. The following list of active ingredients may help you identify these older products. An asterisk (*) indicates the chemical is restricted only in certain forms or concentrations.

**Insecticides**
aldicarb
aldrin
aluminum phosphide
amitraz
azinphos
BHC
bithinol
carbofuran
carbophenothion
chloranil
chlordane
chlordimeform
*chlorfenvinphos
chlorobenzilate
copper arsenate
curacron
cypermethrin
DDT
demeton
diallate
*dicrotophos
dieldrin
diflubenzuron
*dioxathion
*disulfoton

## GARDENING WITHOUT PESTICIDES

Do you really need pesticides—or at least as much as you've been using? Probably not. There is a growing awareness of alternative gardening methods, some using no chemicals at all, others using them selectively. Employing names such as "biological control,"

endrin
*EPN
*ethoprop
ethyl parathion
*fensulfothion
fenvalerate
flucythrinate
*fonofos
heptachlor
kepone
*lindane
*methamidophos
methidathion
*methomyl
methyl parathion
metribuzin
*mevinphos
mirex
milban
*monocrotophos
*permethrin
*phorate
*propetamphos
*sodium arsenate
strobane
sulfotepp
sulfuryl fluoride
TEPP
*terbufos
toxaphene
tralomethrin
triphenyltin hydroxide

**Soil fumigants**
allyl alcohol
chloropicrin
DBCP
*ethoprop
*methyl bromide

**Weed killers**
2,4,5-T
acrolein
*arsenic trioxide
diclofop methyl
nitrofen
*paraquat
*picloram
pronamide
silvex

**Snail and slug poisons**
clonitralid

**Rat, mouse, gopher poisons**
chlorophacinone
phosacetim
sodium fluoroacetate
*strychnine
thallium sulfate

"integrated pest management," "conservation farming," and "genetic manipulation," they fill dozens of magazines and are the subjects of numerous books.

Here are some general guidelines for ways to control pests and weeds without using pesticides:

## *How to Read a Pesticide Label*

Product labels are the Green Consumer's most important source of information about a pesticide's potential effect on the environment and on human health. Labels tell you how to mix and apply a product, which active ingredients it contains, and which pests it is intended to kill. *It is against federal law to use any pesticide in any way not in accordance with instructions on the label.*

The label below shows the information you will find on a typical pesticide label. The key to the highlighted numbers is on the facing page.

### Back or Side Panel

STATEMENT OF PRACTICAL TREATMENT

IF SWALLOWED: Speed is imperative. Call a physician or Poison Control Center. Drink 1 or 2 glasses of water and induce vomiting by touching back of throat with finger. Do not induce vomiting or give anything by mouth to an unconscious person. Apply artificial respiration if breathing stops. IF ON SKIN: Wash thoroughly and immediately with cold running water and/or diluted vinegar. Do not use soap. IF IN EYES: Flush with plenty of water. Get medical attention if irritation persists.

PRECAUTIONARY STATEMENTS

HAZARDS TO HUMANS AND DOMESTIC ANIMALS. WARNING: May be fatal if swallowed, inhaled or absorbed through the skin. Do not breathe vapors or spray mist. Do not get in eyes, on skin, or on clothing. If spilled on clothing, remove and wash clothing before reuse. Keep away from children, domestic animals, and foodstuffs.

ENVIRONMENTAL HAZARDS

This product is toxic to birds and other wildlife. Keep out of lakes, streams, and ponds. Do not apply when weather conditions favor drift from treated areas.

DIRECTIONS FOR USE

It is a violation of Federal Law to use this product in a manner inconsistent with its labeling.

STORAGE: To be stored in original container and placed in areas inaccessible to children. PESTICIDE DISPOSAL: Product remaining in original container should be disposed of by securely wrapping trash. CONTAINER DISPOSAL: Do not reuse empty container. Rinse thoroughly before discarding in trash. NOTICE: Buyer assumes all risks of use, storage, or handling of this product not in strict accordance with directions given herewith.

**DIRECTIONS FOR USE CONTAINED ON ENCLOSED INSTRUCTION SHEET.**

**[Distributor name and address]**

### Front Panel

NAME OF PRODUCT

Garden Spray

9

For use on ornamentals, fruits and vegetables. Controls: Aphids, Leafhoppers, Thrips, and similar sucking insects as specified on the instruction sheet.

8

| ACTIVE INGREDIENT | |
|---|---|
| Nicotine expressed as alkaloid | 40% |
| INERT INGREDIENTS | 60% |
| TOTAL | 100% |

KEEP OUT OF REACH OF CHILDREN

**WARNING**

See rear panel for statement of practical treatment and other precautionary statements

EPA Est. No. 5887-IL-1

EPA Reg. No. 5887-7

Net Contents 2 oz. Avdp.

Prepared by Cheryl Best and the staff of *Garbage* magazine.

## Key to Pesticide Label

1. Directions for first aid in case of overexposure.
2. Cautions regarding use of the pesticide and its human health hazards.
3. A list of all known hazards to wildlife, beneficial insects, groundwater, etc., plus general instructions for avoiding these hazards.
4. Detailed instructions on how to mix and apply the pesticide and a list of all pests and crops for which the EPA permits its use. It is a federal offense to use any pesticide in any fashion not in accordance with these instructions.
5. Name of the company distributing the pesticide.
6. U.S. Environmental Protection Agency registration number (which signifies that the product has been registered with the EPA for the uses stated on the label) and establishment number (which indicates the facility where the pesticide was manufactured).
7. Signal word, listed in large letter, indicates toxicity: CAUTION indicates the product is relatively nontoxic or slightly toxic. WARNING means it is moderately toxic. DANGER/POISON indicates it is highly toxic.
8. Inert ingredients are substances in the formulation that do not kill the pests, listed by percentage only, not by name. Active ingredients are chemicals that kill the pests, listed by name as a percentage of the formulation.
9. The pesticide's common or brand name.

❑ Don't try to force plants to grow in environments that don't suit them. A rose that is grown against a house, for example, where it will be short of water, will inevitably suffer from mildew.
❑ If you know you have a particular pest or disease in your garden, try to find plants that are naturally resistant. Buy certified virus-free stock when planting soft fruit or potatoes. Be wary of accepting plants from other gardeners—they may be the horticultural equivalent of the Trojan Horse.
❑ Don't plant large groups of one type of plant—plant a variety of plants. Gardeners have long believed, for example, that if you plant marigolds alongside carrots or potatoes, you can control pests like carrot fly or eel worm.
❑ Make sure you sow seed or plant seedlings at the right time. Vigorous growth is often a plant's best defense. Plant too early, and your plants will be unnecessarily vulnerable.

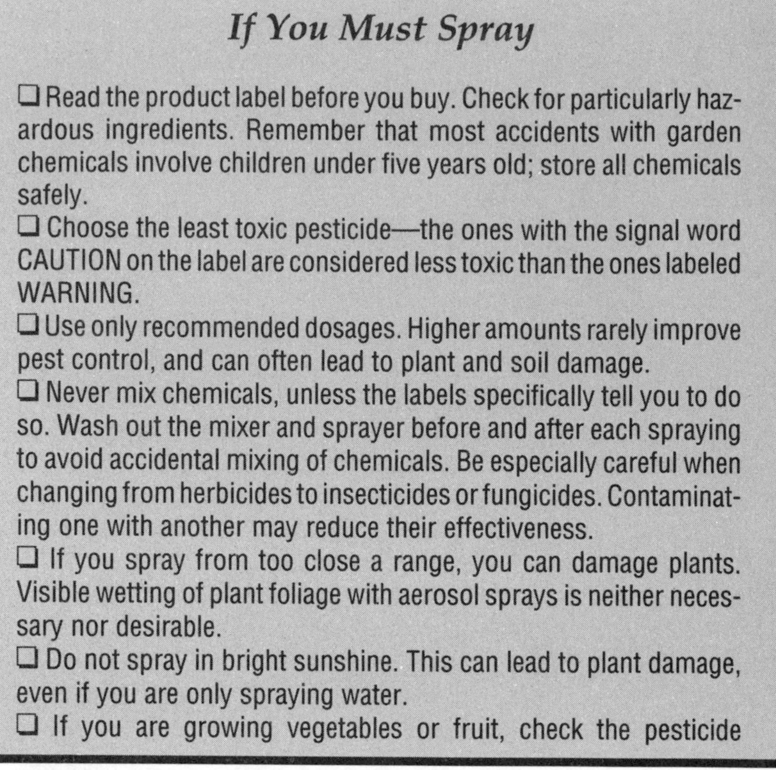

## *If You Must Spray*

❑ Read the product label before you buy. Check for particularly hazardous ingredients. Remember that most accidents with garden chemicals involve children under five years old; store all chemicals safely.
❑ Choose the least toxic pesticide—the ones with the signal word CAUTION on the label are considered less toxic than the ones labeled WARNING.
❑ Use only recommended dosages. Higher amounts rarely improve pest control, and can often lead to plant and soil damage.
❑ Never mix chemicals, unless the labels specifically tell you to do so. Wash out the mixer and sprayer before and after each spraying to avoid accidental mixing of chemicals. Be especially careful when changing from herbicides to insecticides or fungicides. Contaminating one with another may reduce their effectiveness.
❑ If you spray from too close a range, you can damage plants. Visible wetting of plant foliage with aerosol sprays is neither necessary nor desirable.
❑ Do not spray in bright sunshine. This can lead to plant damage, even if you are only spraying water.
❑ If you are growing vegetables or fruit, check the pesticide

❑ Identify your specific problem. The more you know about the pest or disease that is damaging your plants, the less likely it is that you will use the wrong garden chemical. Often you will find that the problem will solve itself. The bean weevil, for example, eats notches out of broad bean leaves, but beans normally grow so fast that there is no danger to the crop.
❑ Pests such as caterpillars and sawfly larvae can often simply be picked off by hand. Diseased leaves, fruit, and other plant material should be cleared away. In this case, you may need to resort to a bonfire to ensure sterilization of material you may wish to return to the garden. The ash can be added to enhance your compost heap.
❑ Whatever you grow, encourage natural pest and disease controls. The best way to do this is to provide habitats and food plants for some of the insects and other creatures that prey on pests. A small pond could provide a home for toads and frogs, for example,

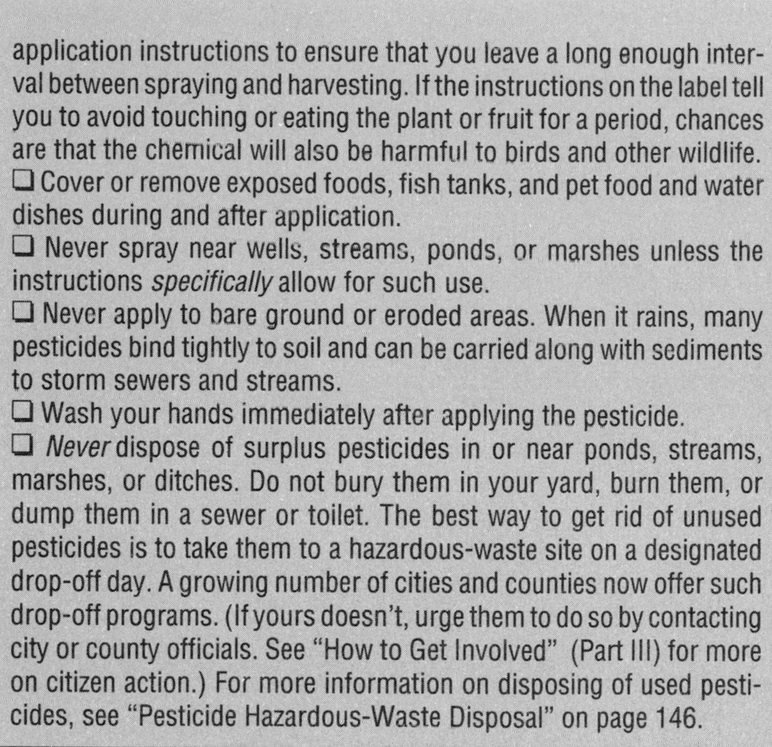

application instructions to ensure that you leave a long enough interval between spraying and harvesting. If the instructions on the label tell you to avoid touching or eating the plant or fruit for a period, chances are that the chemical will also be harmful to birds and other wildlife.

❑ Cover or remove exposed foods, fish tanks, and pet food and water dishes during and after application.

❑ Never spray near wells, streams, ponds, or marshes unless the instructions *specifically* allow for such use.

❑ Never apply to bare ground or eroded areas. When it rains, many pesticides bind tightly to soil and can be carried along with sediments to storm sewers and streams.

❑ Wash your hands immediately after applying the pesticide.

❑ *Never* dispose of surplus pesticides in or near ponds, streams, marshes, or ditches. Do not bury them in your yard, burn them, or dump them in a sewer or toilet. The best way to get rid of unused pesticides is to take them to a hazardous-waste site on a designated drop-off day. A growing number of cities and counties now offer such drop-off programs. (If yours doesn't, urge them to do so by contacting city or county officials. See "How to Get Involved" (Part III) for more on citizen action.) For more information on disposing of used pesticides, see "Pesticide Hazardous-Waste Disposal" on page 146.

whose diet includes slugs. Attract birds by providing nesting sites (bird boxes, hedges, and space behind climbers) and winter food (leave ornamental plants to go to seed and provide food dispensers). Grow hardy annuals like *Limnanthes douglassi* (the "poached egg" plant) and *Convolvulus tricolor*, which provide food for the syrphid fly, whose larvae consume large numbers of greenfly, blackfly, and other aphids. (See "Natural Remedies for Pests and Diseases" below for additional suggestions.)

❑ Protect crops with a physical barrier. If you are a Green Consumer, you probably don't buy plastic soda bottles, but one idea is to collect everyone else's plastic bottles and saw them in half. The ends can be used as mini-cloches for seedlings. Use old carpet padding around the base of your brassicas, which will help confuse the cabbage root fly.

❑ If, after all this, a particular pest gets out of hand and you feel you must use a chemical spray, use as little as possible.

## —Natural Remedies for Pests and Diseases—

**Insecticidal soap:** This natural soap destroys pest membranes. It is effective against aphids, crickets, earwigs, mealybugs, rose slugs, scales, spittlebugs, white flies, and many others. One brand is **Safer's Insecticidal Soap.**

**Bacillus thuringiensis:** BT is a highly selective biologic insecticide that is particularly effective against leaf-eating caterpillars. It kills them by paralyzing the digestive tract. Brands include **Safer's Natural Caterpillar Killer, Dipel 2X, Javelin Liquid BT,** and **Entice Insect Feeding Stimulant**.

**Milky spore:** A natural bacteria that kill the grub phase of Japanese beetles. The milky spores actually remain alive in the soil, preventing new infestations for a few years.

**Dormant oil sprays:** Oil sprays, such as **Safer's Sunspray**, can be used to control scale insects, red spider, mites, mealybugs, and whitefly larvae on azaleas, evergreens, fruit trees, shade trees, shrubs, woody plants, and all other ornamentals.

**Stale beer:** Put this out at night to attract slugs.

**Homemade sprays:** One popular recipe calls for liquefying the following in a blender: 3 large onions, 1 whole garlic, 2 tablespoons hot red pepper in 1 quart of water. Stir in 1 tablespoon of soap. Apply with any spray bottle.

**Insect-eating insects:** These "good" bugs include ladybugs, lacewings, dark ground beetles, soldier beetles, and praying mantises, which will feast on some of your garden pests. These may be available in nurseries.

**Natural Fertilizers.** One form of protection against all sorts of pests and diseases is healthy soil. If you enrich it with the right fertilizers (as well as with humus and compost), your soil will be able to withstand insect attacks with greater success.

Before you fertilize, you may want to have your soil tested.

You can do this yourself by using soil test kits, or have a county extension agent analyze the soil for organic content and acidity. Acid soils can be balanced by applying powdered lime with a spreader.

Avoid artificial fertilizers in favor of organic manures. Artificial fertilizers can make the soil acid and drive worms away, and may also trigger rapid sappy growth in plants by releasing a burst of free nitrate. Ironically, too, overuse of fertilizers can damage overall soil fertility by "locking up" essential elements such as calcium and magnesium and by suppressing the activity of natural soil organisms that normally fix nitrogen from the air or make phosphate available to plants. In addition, excess fertilizers can reach your local stream and lead to water pollution problems.

Avoid applying fertilizer on windy days or just before a heavy rain. For best results, always apply fertilizers according to the directions on the package.

Mineral fertilizers should be used as a supplement to, rather than as a replacement for, natural nutrient recycling in the soil. In general:

**Use**

basic slag
bone meal
calcified seaweed
calcium sulfate
feldspar
fish meal
ground chalk
hoof and horn meals
limestone
magnesium
rock phosphate
rock potash
seaweed
unadulterated seaweed foliar sprays
wood ash

**Restrict the use of**

aluminum phosphate borax
dried blood
Epsom salts
hop waste
kieserite
leather meal
sulfate of potash
wood shoddy

**Avoid all other mineral fertilizers, including**

Chilean nitrate
muriate of potash
nitrochalk
quicklime
slaked lime
urea

The natural and organic garden suppliers listed on pages 154-155 include many sources of worthwhile fertilizers.

## Pesticide Hazardous-Waste Disposal

| Chemical Products | Hazardous Ingredients | Alternatives | Hazard Properties | Disposal |
|---|---|---|---|---|
| Arsenicals | lead arsenate, calcium arsenate, Paris green | live traps, remove food supply | toxic | 1 |
| Botanicals | pyrethrins, rotenone, nicotine | insecticidal soap, import predators (e.g. ,ladybugs, ground beetles, praying mantis) | toxic | 1 |
| Carbamates | carbaryl (Sevin-r), aldicarb (Temik-r), carbofuran (Furadan-r), propoxur (Baygon-r) | keep garden weed-free; import predators, insecticidal soap | toxic | 1 |
| Chlorinated hydrocarbons | DDT, Aldrin, Endrin, kepone, Dieldrin, heptachlor, chlordan, Dicofil, Lindane | keep garden clean, import predators, insecticidal soap | toxic | 1 |
| Organo-phosphates | Parathion, Malathion, Diazinon-r, Dichlorvos, Chlorpyrifos | remove plant debris from garden, insecticidal soap | toxic | 1 |
| Flea collars and sprays | carbamates, pyrethrins, organo-phosphates | herbal collar or ointments or brewer's yeast in pet's diet | toxic | 1 |
| Fungicides | captan, folpet anilazine, zinc/copper compounds | do not over-water, keep areas clean and dry | toxic | 1 |
| Herbicides | 2,4-D, glyphosate, prometon | strong hoeing or hand weeding; keep grass short | toxic | 1 |
| Houseplant insecticide | methoprene, malathion, tetramethrin, carbaryl | mix bar soap and old dishwater; spray on leaves, rinse | toxic | 1 |
| Rat and mouse poisons | brodifacoum, coumarins (e.g., warfarin), strychnine | live traps, remove food supply | toxic | 1 |
| Roach and ant killers | organo-phosphates, carbamates, pyrethrins | roaches: traps or baking soda and sugar mix; ants: chili powder | toxic | 1 |

1. These wastes should be safely stored until a hazardous-waste program is organized in your community. Fully spent currently available pesticide containers may, however, be triple rinsed and the rinse water poured down the drain and flushed with water.

Courtesy Environmental Hazards Management Institute, 10 Newmarket Rd., P.O. Box 932, Durham, NH 03824; 603-868-1496

## Healthy Lawns, Healthy People

It is simply astounding how many chemicals Americans dump on their lawns in the never-ending quest for the immaculate, perfectly manicured, crabgrass-and-dandelion-free grass. Whether you apply these chemicals yourself or hire a gardener or lawn care company to do it, it's likely that the quest for the perfect lawn will inevitably lead you to apply one or more dangerous, unhealthy substances to your lawn.

Lawn care chemicals can make you—and your neighbors, your children, and your pets—sick. Most of the pesticides used in controlling weeds and insects are broad-spectrum biocides, which means they are poisonous to a wide variety of living organisms, including garden plants, wildlife, pets—and people. Inert ingredients, which may constitute 50 to 99 percent of a pesticide formula, may actually be more toxic than the active ingredients. The poisons can be absorbed through the skin, by the mouth, or by the breathing in of sprays, dusts, or vapors. You can be poisoned if you apply or are present during application; touch contaminated grass, shoes, clothing, or lawn furniture; or put contaminated objects (toys, golf balls) or fingers in your mouth. Moreover, the chemicals aren't necessarily safe once they dry. They can remain active for months, during which time they can release toxic vapors. Breathing these vapors, even from neighbors' lawns or while playing on or mowing contaminated grass, can make you sick.

The symptoms of lawn pesticide poisoning are deceptively simple and, unfortunately, similar to those of many other illnesses. Simply put, pesticides attack the central nervous system and other vital body centers. Symptoms include sore nose, tongue, or throat; burning skin or ears; skin rashes; excessive sweating or salivation; chest tightness; asthmalike wheezing attacks; coughing; muscle pain; headaches or eye pain; cramps; and diarrhea. Even harder to detect are some of the long-term problems, including lower male fertility, miscarriage, birth defects, and liver and kidney dysfunction.

Other frequent victims are birds, especially songbirds. What wildlife specialists have come to call the "lawn care syndrome" in birds refers to the classic signs of pesticide ingestion: shivering, excessive salivation, grand mal seizures, wild flapping, and sometimes screaming, according to one U.S. Fish and Wildlife Service wildlife toxicologist. While chemical-related deaths of bald eagles and other protected species have long been documented, songbird poisoning from lawn care products is a little-studied environmental problem, one that began to surface only with the growth of gardening chemicals in the 1980s.

One of the biggest sources of lawn care chemicals are professional lawn care companies. According to *Lawn Care Industry*, a trade magazine, there are an estimated 5,000 companies that groom, mow, and spray more than 7 million lawns a year with an estimated 8 million pounds of pesticides—at a price tag to consumers of about $1.5 billion, nearly double that of a decade ago. But the actual price has been higher: Many of the chemicals used by these companies have posed a serious health threat. According to a study by Public Citizen of the 40 most commonly used lawn care pesticides, 9 were classified by the Environmental Protection Agency as "possible" or "probable" human carcinogens. Two of the herbicides, **chlorothalinol,** which sells under the trade name **Daconil,** and **diazinon** have been linked with the deaths of two golfers who died after playing on courses treated with these chemicals. (The government banned diazinon for use on golf courses in 1988, but says it does not have enough data on the long-term health effects of chlorothalinol to ban its use.) Public Citizen found one of the most widely used compounds in lawn care to be **2,4-D**, an ingredient in the defoliant Agent Orange. In 1986, the National Cancer Institute linked 2,4-D to an increased cancer risk among the farmers who used it.

Despite this, many of the lawn care companies do little to educate their customers to the potential risks. Public Citizen surveyed company literature and found several companies saying just the opposite. For example, **Orkin Lawn Care** of Atlanta stated that its "fully approved and tested chemicals are completely child- and pet-safe." But says Public Citizen's pesticide specialist Laura Weiss, "To call any pesticide mixture 'safe' is inaccurate. Pesticides, by definition, are compounds that kill, and all chemicals used in lawn care service pose some risk to human health. The level of risk depends on a variety of factors, including the toxicity of the chemical, the extent of exposure, and a person's sensitivity to the chemical. Children are particularly vulnerable to increased cancer risks because their cells develop more rapidly than the cells of adults, and their play habits are likely to bring them into close contact with lawn care chemicals." (Copies of *Keep Off the Grass*, the Public Citizen study, are available for $8 from Public Citizen's Congress Watch, 215 Pennsylvania Ave. SE, Washington, DC 20003; 202-546-4996.)

Not all lawn care services are bad. A growing number use natural or organic methods, and others will consider your concerns about pesticides in their treatment practices. **ChemLawn Services Corporation**, one of the national chains that formerly used 2,4-D (it stopped using the chemical in 1987), now offers a fertilizer-only package and is testing an organic lawn care program. For customers concerned about chemicals used in their lawn care, a ChemLawn spokesperson said it will "work with the customer" to find acceptable alternatives.

**Solutions for Green Lawns.** As with other aspects of gardening, a healthy lawn is its own best defense against weeds and pests. It will crowd out most weeds and resist insects and disease.

The three essentials of a healthy lawn, according to Stuart Franklin, author *Building a Healthy Lawn* ($9.95 plus $2.50 for shipping from Garden Way Publishing, Schoolhouse Road, Pownal, VT 05261; 802-823-5811, 800-441-5700), are the roots, the top growth (the grass blades), and the soil. They all interact to give a lawn its level of health—or illness. The roots must go deep into the soil so they can seek out soil nutrients and find water during dry periods. Short mowing and light watering will prevent roots from going deep. Shallow-rooted lawns are the first to die or weaken during the summer heat. In this condition they attract insect and disease organisms.

Franklin, who owns and operates Naturalawn lawn care in Buffalo, New York, suggests the following steps to a healthy lawn:

### In the fall

❑ Fertilize. Grass will best use fertilizer in the fall.
❑ Rake leaves and add to a new or established compost pile for ready fertilizer in the spring.
❑ Service and store mowing equipment properly for trouble-free start-up in the spring.

*[continued on page 152]*

| Pesticide | Chemical Family | Primary Uses | Human Effects | Environmental Effects |
|---|---|---|---|---|
| ROTENONE (NOXFISH ™ PRENTOX™) | Botanical | Insecticide used primarily on vege table crops for various beetles and cabbage worms | Slows breathing and heart rate; irritating to eyes | Breaks down quickly on plants and soil; lasts up to one month in water. Highly toxic to fish; low toxicity to birds and honeybees |
| NICOTINE SULFATE (BLACKLEAF 40™) | Botanical | Kills many sucking insects and larvae | Injures the nervous system; readily absorbed through skin | Breaks down quickly kills mammals, birds fish, and beneficial insects; relatively safe to bees |
| Propoxur (Baygon™) | Carbamate | Insecticide for cockroaches, flies, and other household pests as well as lawn pests | Probably a carcinogen; temporarily disrupts function of the nervous system | Lasts up to one month on plants; toxic to fish and other wildlife; hazardous to honeybees; some formulations restricted |
| Chlorpyrifos (Dursban™, Lorsban™) | Organo-phosphate | Insecticide for lawn pests, some house pests, ticks, and mosquitoes | Disrupts function of the nervous system | Persists up to one year in soil; toxic to bees, fish, birds, and other wildlife; many formulations restricted |
| Diazinon (Spectracide™, KnoxOut™) | Organo-phosphate | Controls many insects and mites on fruits, vege-tables, ornamentals, and lawns; also used on insects | Disrupts function of the nervous system | Lasts 1-2 weeks on plants; 2 months in soil; toxic to bees, fish, waterfowl, and other wildlife |
| 2,4-D (Weed-B-Gon™, Weed & Feed™) | Chlorophenoxy | The most popular home weed killer | Irritating to lungs, stomach, and intesti-nal linings; can injure liver, kidneys, and nervous system | Persists in water and soil up to 6 months; toxic to bees and fish |
| Carbaryl (Sevin™) | Carbamate | Insecticide used on many fruit, vegetable, tree, turf, pet, and flowering plant pests | Chronic exposure causes loss of appetite and weight, and weakness | Lasts several weeks on plants; 3 months in soil; highly toxic to honeybees; slight hazard to birds, fish, and beneficial insects |
| RYANIA | Botanical | Controls European corn borer, codling moth, cranberry fruit worm, and oriental fruit flies on fruit and some vegetable crops | Relatively nontoxic | Lasts longer than the other botanicals in the environment relatively safe to wildlife and bene-ficial insects |

| Pesticide | Chemical Family | Primary Uses | Human Effects | Environmental Effects |
|---|---|---|---|---|
| Malathion (Cythion™ for indoor use) | Organo-phosphate | Controls many chewing and sucking insects on fruit, vegetables, trees, ornamentals, and houseplants; also controls household pests | Disrupts function of the nervous system | Does not persist in long in the environnment; toxic to fish, honeybees, and beneficial insects; low toxicity to mammals |
| PYRETHRUM (PYRENONE™, PYROCIDE™) | Botanical | Controls flies, mos quitoes, and fruit and vegetable pests, includig aphids, many cabbage loopers, leaf hoppers, white fly, and corn earworm | Relatively nontoxic | Lasts longer than most botanicals in the environment; toxic to honeybees, many beneficial insects, fish, and other cold blooded animals |
| Glyphosate (Kleenup™, Roundup™) | Phosphonomethyl glycine | Used to clear brush, control grassy weeds and poison ivy, and for spot treatments on lawns | Irritates lungs and eyes | Quickly inactivated in soil; low to moderate toxicity to fish and wildlife |
| Methoxychlor (Marlate™) | Organochlorine | Controls mosquitoes flies, ticks, weevils, leafhoppers, Japanese beetles, and other insects; not effective against aphids and mites | Disrupts nervous system; some buildup in fat cells; readily absorbed through skin | Lasts several weeks on plants and longer in soil and water; toxic to fish and bees, little or no hazard to birds, beneficial insects, and mammals |
| SABADILLA (RED DEVIL DUST™) | Botanical | Insecticide for vegetable-crop pests, including striped cucumber beetles, squash-bugs, leafhoppers, and European corn borer | Relatively nontoxic; irritates mucous membranes of nose and throat | Relatively safe to wildlife, but toxic to honeybees |
| Benomyl (Benlate™, Tersan 1991™) | Dithiocarbamate | Fungicide for use on fruit, vegetables, ornamentals, and lawns to control many fungus dieases | Possible carcinogen; mildly irritating to skin, eyes, nose, and throat | Persistent in the environment; hazardous to fish; relatively nonhazar dous to honeybees; increases in toxicity as it breaks down |

Names in all CAPITALS indicate natural pesticides. All others are artificial chemical pesticides.

Prepared by Cheryl Best and the staff of *Garbage* magazine.

## *Gardening Without Water*

Can you garden with little or no water? It may not seem possible—water is the lifeblood for most plants—but a growing number of people are finding that "dry" landscaping makes sense.

The idea—called "xeriscaping" (from the Greek root *xeros*, or "dry")—is catching on in water-starved parts of the country, particularly in the West, where water rationing is an all-too-frequent necessity. A xeriscaper's garden is likely to include many of the drought-tolerant plants found across the Southwest and other hot, dry climates. Seed companies specializing in xeriscaping even offer low-water-use grass seeds. Buffalo grass and blue gamma, two grasses popular in Colorado, for example, are generally richly green through the end of June, then turn greenish-yellow for the rest of the summer, then return to green in the spring. Yet these lawns generally need no watering other than the sparse summer rain showers. (However, with a little extra water, they will stay green longer.) Not all of these grasses will grow in every part of the country, however. Check with a local nursery for information about what grows in your area.

Several seed companies now offer xeriscaping seeds and devices to minimize water use. Even the catalog from the giant **W. Atlee Burpee Co.** offers several drought-resistant flowers. Here is a selected list of companies. Unless otherwise indicated, each offers a free product catalog.

❑ Purchase seed and limestone as needed for late winter application.

### In the spring

❑ Don't be overanxious to start gardening. Stay off the lawn if it is partly frozen or soggy.

❑ Gently rake off leaves and debris to prevent fungus disease and dead spots.

❑ Rake matted grass so it is standing straight up. Allow air and light to penetrate.

❑ Make your first cuts short to clear off dead blades and encourage the grass to spread. Then raise your cutting height as the grass begins to grow vigorously.

Bernardo Beach Native Plants
One Sanchez Rd.
Veguita, NM 87062
505-345-6248
*Send four 25-cent stamps for catalog.*

Botanicals
219 Concord Rd.
Wayland, MA 01778
508-358-4846
*Catalog $1.50.*

W. Atlee Burpee Co.
300 Park Ave.
Warminster, PA 18974
215-674-4915, 800-888-1447
*Free catalog.*

Old Farm Nursery
5550 Indiana St.
Golden, CO 80403
303-278-0754
*Free catalog.*

Plants of the Southwest
930 Baca
Santa Fe, NM 87501
505-983-1548
*Catalog $1.50.*

Rain Bird Irrigation
145 N. Grand Ave.
Glendora, CA 91740
818-963-9311, 800-247-3782
*Free catalog.*

Another helpful reference is *Waterwise Gardening*, $6.95, from **Sunset Books** (80 Willow Rd., Menlo Park, CA 94025; 415-331-3600, 800-227-7346). Add $1.25 postage and handling.

❑ Don't cut off more than one-third of the blade at once; it's a shock to the plant.
❑ Seed bare spots and overseed thin spots by mid-spring. If grass isn't there, weeds surely will be.
❑ Fertilize early if you didn't do a late fall fertilizing. You want to get the grass tall and thick before weeds (especially crab grass) begin to sprout.
❑ Keep your blades *sharp*. Don't tear the grass. Cut it.
❑ Try using an organic fertilizer. Put some life back into your soil. Organic fertilizers and other organic supplements will also help to break up thatch.
❑ Leave short clippings on the lawn unless you have a thatch problem. Clippings won't normally cause thatch, but they will add to it once it is there.

## Low-Impact Garden Supply Companies

With low-impact gardening (using minimal chemicals) or organic gardening (using no chemicals), the best place to start is with the seed. Several seed suppliers offer low- or no-chemical seeds. For example, at **Johnny's Selected Seeds** (Foss Hill Rd., Albion, ME 04910; 207-437-9294), seeds are grown with minimal pesticide use. **The Sprout House** (40 Railroad St., Great Barrington, MA 01230; 413-528-5200) offers organic, chemical-free seeds. For herb seeds raised with botanical pest control, contact **Meadowbrook Herb Garden** (Route 138, Wyoming, RI 02898; 401-539-7603).

Good examples of natural pest control products are those manufactured by **Safer, Inc.** (189 Wells Ave., Newton, MA 02159; 617-964-2990), including lawn, garden, and pet products made with nontoxic ingredients that biodegrade rapidly and leave no residues. Most Safer products fall into the EPA's "least hazardous" pesticide category. Safer products, available in stores and through many mail-order catalogs, include house plant products (such as insecticidal soap), garden products, soil and turf insecticides, herbicides, and insect traps.

Below is a list of select low-impact garden supply sources. Many offer a comprehensive range of lawn and garden care products; others do not. Be sure to inquire for the products you are interested in before asking for a catalog.

Gardens Alive!
P.O. Box 149
Sunman, IN 47041
812-623-3800

Growing Naturally
P.O. Box 54
Pineville, PA 18946
215-598-7025

Green Pro
380 S. Franklin St.
Hempstead, NY 11550
516-538-6444, 800-645-6464

Integrated Fertility Management
333 Ohme Gardens Rd.
Wenatchee, WA 98801
509-662-3179, 800-332-3179

Meadowbrook Herb Garden
Route 138
Wyoming, RI 02898
401-539-7603

Mellinger's
2310 W. South Range Rd.
North Lima, OH 44452
216-549-9861, 800-321-7444

Nature's Control
P.O. Box 35
Medford, OR 97501
503-899-8318

Nitron Industries, Inc.
4605 Johnson Rd.
P.O. Box 1447
Fayetteville, AR 72702
501-750-1777

The Necessary Trading Co.
New Castle, VA 24127
703-864-5103

Ohio Earth Food, Inc.
13737 Duquette Ave. NE
Hartville, OH 44632
216-877-9356

Ringer
9959 Valley View Rd.
Eden Prairie, MN 53444
612-941-4180, 800-654-1047

Universal Diatomes, Inc.
410 12th St. NW
Albuquerque, NM 87102
505-247-3999

Pheromones, the natural sex-attractants of insects, are being applied to traps and used as pest control devices. Two companies offering pheromone trapping systems are:

Consep Membranes Inc.
213 SW Columbia St.
P.O. Box 6059
Bend, OR 97708
503-388-3688

Insects Limited, Inc.
10540 Jessup Blvd.
Indianapolis, IN 46280
317-846-3399, 800-992-1991

## Composting

Composting is the process by which organic wastes—including food wastes, paper, and yard wastes—can decompose naturally, resulting in a product rich in minerals. Gardeners love compost—it makes an ideal soil conditioner, mulch, resurfacing material, or landfill cover—and so do plants.

During composting, a mass of biodegradable waste, combined with sufficient moisture and oxygen, "self-heat"—a process by which microorganisms metabolize into organic matter and release energy in the form of heat. The process is nothing more than an accelerated version of the breakdown of organic matter that occurs under natural conditions, such as on the forest floor. Because composting is a natural process, it can be carried out with

## *Plastics in the Garden*

Plastics seem to be everywhere—even in the garden. A growing number of plastic products have been marketed to enhance gardening. Plastic products in the garden include mulches, row covers or floating blankets of spun-boned polymers, support netting for pea and bean trellises, shade netting for protection of cool-weather plants, irrigation equipment, pots, trays, and markers and ties. Like the plastics used in everything from food packaging to fiberfills, these plastics can wreak havoc on the environment.

Fortunately, there are nonplastic alternatives to all of these products; your garden supply dealer can point you in the right direction, or you may contact one of the mail-order companies listed on pages 154–155.

If you must purchase these plastic products, handle them carefully so that they last as long as possible. If you cut fabrics to fit a specific area, mark it well so it can be easily reused the following season. Drain hoses before storing for the winter so they won't freeze and split. When not in use, store plastics away from sunlight.

Many plastic garden products can be reused for other purposes. For example, plastic seedling pots can become hot caps to protect small seedlings from frost. Worn plastic hoses can tie back heavy limbs or may become supports for newly planted saplings.

While you're in the process of reusing plastic, check to see what other household items might be of use in the garden. Plastic shopping bags can covered with shredded bark or leaves for weed control. Empty milk and water jugs can protect tender young plants.

as little, or as much, intervention and attention as the composter desires.

While several million Americans have ongoing compost piles in their yards, the growing action is going on in communities, which are finding that large-scale composting of leaves and other yard waste not only reduces solid waste but actually can create a marketable—and profitable—final product. About 1,000 cities and counties now have leaf-composting facilities, according to the Environmental Protection Agency. Communities variously give the compost away or sell it to residents, use it for public park service projects, sell it to other communities, or trade it for nursery stock. With yard wastes constituting about a fifth of all household waste, this material is a continuously renewable resource.

Examples abound of communities turning their garden trash into cash. For more than a quarter century, for example, the borough of Swarthmore, Pennsylvania, has collected about 400 truckloads of leaves annually, which they spread over a vacant lot owned by Swarthmore College. Each August, the leaves are bulldozed into a large pile, adjacent to piles from three prior years. Allowed to decompose over a two- to three-year period, the borough is left with nearly 100 loads of rich leaf mold in the oldest pile. That, in turn, is sold to residents for $25 per truckload delivered.

Starting your own backyard compost heap is easy. Simply gather fallen leaves, dead plants, and brush in a corner of your backyard. Bacteria, fungi, and other organisms will break it down, eventually creating compost, a loose, crumbly earth. The key is to achieve a proper balance of moisture and oxygen. Experts say to keep the pile under five feet tall and to keep it moist, but not wet. Layer the pile using green leaves (which supply oxygen) and brown leaves or straw (which supply carbon).

Whether you compost or not, don't collect yard wastes in plastic bags. A better alternative is the **Ecolobag** ($30 for 50 bags from **Dano Enterprises, Inc.,** 75 Commercial St., Plainview, NY 11803; 516-349-7300), a heavy-duty, weather-resistant 50-pound kraft paper bag specifically designed for collecting leaves and other biodegradable trash. The convenient bag has a 12-by-16-inch square bottom that allows it to stand by itself.

## How Trees Save the Earth

You may think of trees as a gift from Mother Nature, something to climb up or sit under, or the cause of fall raking. You probably don't think of trees as life savers. But that's exactly what they are. In cities, countryside, and forests, trees breathe life into our planet and save it from a host of environmental problems.

What exactly do trees do? Aside from their beauty and the food some of them produce, some of their other gifts include:

❑ **Cooling cities.** Urban areas are "heat islands"—their buildings, streets, cars, and other structures and activities soak up heat on a summer's day and release it at night. Researchers at the Lawrence Berkeley Laboratory in California found that average temperatures in the city can be five to nine degrees higher than those in surrounding suburbs. Groups of trees can offset this heat, operating as nature's air conditioners. Trees also help to reduce noise in cities.

❑ **Battling the greenhouse effect.** Because they absorb carbon dioxide, trees and other greenery offer the cheapest way to combat the greenhouse effect. (Carbon dioxide is responsible for about half of the greenhouse problem.) The average tree absorbs between 26 and 48 pounds of carbon dioxide a year, according to the American Forestry Association. An acre of trees takes in about 2.6 tons of carbon dioxide, enough to offset the emissions produced by a car driving 26,000 miles.

❑ **Preventing erosion.** Trees protect against the erosive power of wind, helping to protect topsoil and retain soil moisture. Deprived of their protective tree cover, hillsides are easily eroded and less able to retain rainfall. Without trees to break its force, the wind finds the exposed topsoil easy pickings. Continued wind produces giant, gritty clouds that steadily diminish America's precious soil heritage.

❑ **Reducing energy needs.** The shade provided by trees can save considerable energy and money. In the summer, three well-placed trees around a house can cut home air conditioning energy needs by 10 to 15 percent.

With these benefits in mind, several local and national organizations have created ambitious campaigns to plant millions of new trees across the United States as well as in other parts of the world:

❑ **The Basic Foundation** (P.O. Box 47012, Saint Petersburg, FL 33743; 813-526-9562), has created a campaign in cooperation with **Rainforest Action Network** and **Arborfilia**, a nonprofit Costa Rican organization, to plant endangered tropical trees in Costa Rica. Tree planting takes place in community nurseries of small farmers that are trained in the reproduction, planting, and care of endangered native timber trees as well as high-quality fruit trees.

## *Merry (Recycled) Christmas*

What do you do with your Christmas tree after Christmas? If you're like most of the 35 million tree owners each year, you haul it to the trash, where it will end up in a landfill. In the process you are also tossing out a variety of benefits the tree can offer after the holidays.

For one thing, the boughs, cut into one- to two-foot lengths, can serve as a kind of blanket against harsh winter winds for delicate perennials such as azaleas, camelias, and rosebushes. The pleasant-smelling needles and cones can give a piney aroma to a compost heap. Although they take somewhat longer than other foliage to decompose, evergreens will eventually contribute to a rich mixture of compost.

Pine and fir needles also make good soil conditioners, loosening and lightening the consistency of sandy or clay soils. The acidic residue they give off when they decompose is ideal for many plants, including azaleas and rhododendrons. Because decomposing needles can deplete the soil of nitrogen, add some nitrogen to the soil as well.

The tree trunk can be useful in the garden, too, although you'll have to run it through a shredder or chipper. The finished product can be added to the soil as mulch or compost.

Be careful about burning evergreens in your fireplace. The heavy concentrations of resin increase the risk of a chimney fire. Besides, the wood burns too quickly to make for a good fire.

If you aren't willing to take the time to put your old Christmas tree to such uses, try to find someone who is. Several nurseries will accept your tree, which they turn into mulch or wood chips.

In return for the seedlings and technical assistance, the farmers reforest degraded watersheds and areas not suitable for other agriculture. The Basic Foundation accepts tax-deductible contributions: $5 pays for a single tree; $250 for planting a hectare—approximately 1,000 trees.

❑ **TreePeople** (12601 Mulholland Dr., Beverly Hills, CA 90210; 213-273-8733) is a nonprofit group whose stated purpose is "to encourage individuals by education, example, and assistance to improve the environment by planting and caring for trees." TreePeople conducts several programs aimed at both children and adults and maintains a 45-acre "wilderness park" in the heart of

Los Angeles, which includes displays, hiking trails, and a small retail seedling nursery. The organization's book, *A Planter's Guide to the Urban Forest*, is available for a $12 donation. Membership ($25 a year) includes the bimonthly *Seedling News* and six free seedlings each year.

❑ **Trees for Life** (1103 Jefferson, Wichita, KS 67203; 316-263-7294) is responsible for having planted nearly a million trees in India. The group also is trying to organize a program in which the United States and the Soviet Union would help to plant 100 million food-bearing trees in underdeveloped countries. The group seeks donations from individuals and companies.

❑ The **American Forestry Association** (1516 P St. NW, Washington, DC 20005; 202-667-3300) has created a program called "Global ReLeaf," whose aim is to plant 100 million new trees by 1992 to help reverse the greenhouse effect and slow global warming. Several cities have made significant commitments to the campaign; Los Angeles alone has set a planting goal of 5 million trees. According to AFA, there are about 100 million "planting spaces" in the United States where trees could shade homes and businesses. According to Lawrence Berkeley Laboratory, planting those 100 million trees could save American consumers an annual *$4 billion* in energy costs alone. Contact AFA to find out if there is a Global ReLeaf program in your area, or how to set one up if there is not.

❑ **The National Arbor Day Foundation** (100 Arbor Ave., Nebraska City, NE 68410; 402-474-5655) has a campaign to promote tree planting in the United States. The nonprofit organization offers ten Colorado blue spruce trees (six to ten inches tall each), with planting instructions and a copy of *The Tree Book*, with each $10 membership contribution. (People in southern and West Coast states receive ten bald cypress trees, conifers better suited to those climates.)

## Pets and Pet Supplies

Just like their human owners, household pets are susceptible to a wide range of chemicals and other pollutants. It is important to consider ingredients carefully when buying everything from food to flea collars.

It is also important to consider the pets themselves. First and foremost is that you shouldn't buy a dog, cat, bird, fish, or anything else unless you are prepared to take care of it. Each year millions of pets are abandoned—left to roam the streets, starved to death, even flushed down the toilet—by individuals who could not handle the responsibility of having a pet. It is a sickening waste of animal life.

Be particularly careful when choosing an exotic pet, such as tropical fish and turtles. Consider turtles. The Red-Eared Slider, for example, a small baby turtle, was a popular children's pet commonly sold in five-and-dime stores until the mid-1970s. During that time, they were "farmed" in Louisiana in uncontrolled conditions and exported by the millions. Of the 13 million shipped annually, 11 million died en route, or soon thereafter. Government officials eventually outlawed the sale of turtles under six inches.

Abuse against pet turtles may have stopped, but when it comes to tropical birds, it remains rampant. According to the **Humane Society of the United States**, every year millions of birds—20 million in Senegal alone—are trapped for cage bird markets in the United States, Europe, and Japan. Says the Humane Society, "Half of them die before leaving their native soil. The trauma of capture, brutal mishandling, hunger and thirst, gross overcrowding, and disease caused by stress claims a cruel toll of victims. More die in the holds of airlines, in quarantine stations of importing countries, and in pet stores." The society strongly urges individuals to not buy tropical birds for pets.

## Dealing with Pets' Pests

One of the biggest sources of problems for pets and pet owners are fleas and ticks. Fleas are a particular problem. Not only are they a source of torment to dogs and cats, but they can lead to anemia, allergies, skin infections, and tapeworm infections.

One thing to keep in mind is that healthy animals usually don't get fleas in the first place, or at least they get fewer of them. Fleas tend to invade diseased or malnourished animals. Proper diet, adequate water, exercise, sunlight, and social contact all help. Some people feed their pets brewer's yeast, which besides being rich in B vitamins is thought by some veterinarians to repel fleas. One "prescription" calls for feeding two tablespoons of brewer's yeast and a clove of garlic to each animal daily.

But sometimes fleas are inevitable. When they do attack, one of the first steps is a thorough house cleaning: washing all pet bedding in hot water (if the infestation is severe, you may need to destroy the bedding); vacuuming the carpets; mopping the floors; shampooing or vacuuming upholstered furniture. Keep in mind that fleas can be carried through a house by people, too—on their clothing usually—so you may have to clean *everything*, literally. When you're finished, remove the vacuum cleaner bag or other accumulated trash from the house immediately.

You may be forced to apply an insecticide—either by spraying, dipping, or with a medicated shampoo. Many groomers and veterinarians prefer dips because they offer long-term protection, but many dips can be irritating to an animal's skin, especially if used repeatedly. If you've shampooed your animal with a product intended for flea control, it might be better to spray furniture and carpets rather than the animal itself. Pump sprays can direct the insecticide to where it is likely needed, as opposed to "bombs," which coat everything in the house with insecticide. Read directions carefully, however, to ensure that you don't harm upholstery or furniture. Even worse, when several insecticides are used at once, the combination may be deadly—not only to insects, but also to animals. Be sure to follow product directions carefully.

## Insecticide Alternatives

There are several alternatives to spraying or bombing that don't use pesticides. For example:

❑ Aromatic herbs in powder form can be sprinkled over the animal's fur to repel fleas. Herbal flea powders may contain bay leaf, eucalyptus, sage, sassafras, tobacco, or vetiver.

❑ Flea traps, some of them homemade, can be another effective means of flea control. A homemade trap can be built with a shallow plate filled with soapy water. A low-wattage light bulb is aimed toward the plate. Fleas will jump toward the light, land in the dish, and drown.

On page 164 are names and addresses for five of the many manufacturers and distributors of alternative, nontoxic (to animals) pet pest-control products. You may contact each for a catalog.

## *Don't Poison Your Pets*

Gardens are one of the favorite foods of snails during the spring. And snail poison is one of the favorite foods of household pets during that same period. The sweet flavor and foodlike appearance of snail bait is attractive to pets. But the bait can be as poisonous to dogs and cats (and most other small animals) as it is to snails. As an alternative to snail bait, you might try putting a wire mesh fence made of hardware cloth around the garden; hardware cloth is available at most hardware stores. Plant it so it sticks up a few inches above the ground, then curve the top out, away from the garden. It is difficult even for a snail to walk on wire upside down.

Snail bait isn't the only yard and garden hazard for pets. Mole and gopher killers contain strychnine, a sure animal killer. Planting pop bottles at gopher hole entrances (the animals don't like the sound of the wind whistling over the bottle tops) or placing toy windmills in the same area (the vibrations that the windmills transmit to the ground disturb the pests) is a safer and more interesting way to encourage gophers and moles to move on. Chemical herbicides used on dandelions and other weeds can prove troublesome, too, should your pet feast on the treated grass—or even lick its paws after playing in it. To protect your pet's health, always store insecticide sprays and powders out of harm's way, and use them with caution.

If you suspect your pet has accidentally ingested yard and garden poisons, symptoms such as vomiting, trembling, convulsions, dilated pupils, salivation, labored breathing, weakness, or collapse will generally follow. If this happens, try to determine what the poison was, and when and how much was ingested—and call your veterinarian immediately.

By the way, animal pesticides are harmful to humans, too. In 1988, illnesses among workers at a Georgia veterinary hospital prompted federal health officials to warn against excessive exposure to pesticides, particularly **fenthion,** the one the Georgia clinic used to kill fleas. Four workers at the clinic complained of symptoms, including shooting pain, muscle weakness, and numbness. Chronic exposure to pesticides such as fenthion has been shown to cause nerve damage, according to the Centers for Disease Control.

❑ **Baubiologie Hardware** (207B 16th St., Pacific Grove, CA 93950; 408-372-6826) offers several nontoxic pest control products, including "Natural Animal Gentle Dragon Dewormer," "Natural Animal Flea Collar," "Natural Animal Spritz" (for flea and tick control), and "Natural Animal Pet Shampoo Concentrate." Also available are Green Ban products (see below).

❑ **Green Ban** (Box 146, Norway, IA 52318; 319-227-7996) sells pet products made of natural ingredients and completely biodegradable. Products include flea powder and dog shampoo; also available are products for plants and people.

❑ **Necessary Trading Co.** (Newcastle, VA 24127; 703-864-5103) sells several products, including a "Healthy Pet Care Kit," a year's supply of healthy nontoxic control against fleas, ticks, and lice ($17.75).

❑ **Pet Guard** (P.O. Box 728, Orange Park, FL 32073; 800-874-3221, 800-331-7527 in Florida) sells a wide range of natural pet products, including flea powders and shampoos.

❑ **Ringer** (9959 Valley View Rd., Eden Prairie, MN 55344; 612-941-4180, 800-654-1047) sells "Flea & Tick Attack" ($12.98) a natural insect repellent. Ringer also sells many nontoxic garden pest control products.

❑ **Safer, Inc.** (189 Wells Ave., Newton, MA 02159; 617-964-2990) offers a line of products, including "Pet Odor Eliminator," "Flea Soap for Cats & Dogs," and "Indoor Flea Guard," all with natural insect-fighting ingredients.

❑ **Whole Earth Access Company** (2990 Seventh St., Berkeley, CA 94710; 415-845-3000, 800-845-2000) offers several natural pet products, including cedar pet pillows (fleas and ticks hate cedar), in a variety of sizes ($11 to $30). Catalog is $7; $5 are deductible from a $25 purchase.

# —G·R·E·E·N G·I·F·T·S—

There are ideas for "green" gifts throughout this book—a mesh shopping bag in the "Food and Groceries" chapter might make a great present, for example. Below are some environment-minded gifts that will allow you to make a social statement at the same time you make a personal one. Moreover, many of the products below directly benefit environmental organizations and causes. All of these organizations sell their products by mail. Many offer a catalog of their products.

## Clothing

**Environmental Awareness Products** (3600 Goodwin Rd., Ionia, MI 48846; 517-647-2535) distributes Earth Island Institute and Jim Morris Environmental T-shirts. EAP also offers its own "Gaia Expressions" clothing, emphasizing nature, conservation, and Native American themes. A portion of proceeds from sales go to the Earth Island Institute and various other environmental groups.

**Environmental Gifts** (P.O. Box 222-C, Helena, MT 59624; 406-458-6466) sells "Love Your Mother" T-shirts, posters, and window decals.

**Jim Morris Environmental T-Shirts** (P.O. Box 831, Boulder, CO 80306; 303-444-6430) offers a wide variety of colorful T-shirts and sweatshirts with environmental and conservationist motifs. Environmental books from a variety of publishes are also available. At least 10 percent of profits is donated to environmental groups.

**Wearable Arts** (26 Medway, #12, San Rafael, CA 94901; 415-456-7034, 800-635-2781) produces a variety of colorful T-shirts portraying environmental themes, from North America's hummingbirds to Pacific gray whales to tropical rain forests. All T-shirts are 100 percent cotton and are printed with nontoxic inks. Also available are canvas tote bags, ideal for carrying groceries, books, or picnics. Profits from sales are given to the Earth Island Institute.

## Games and Educational Products

**Animal Town Game Co.** (P.O. Box 2002, Santa Barbara, CA 93120; 805-682-7343) offers entertaining games focusing on ecology and the environment. Selections include: "Dam Builders" (age 8 and up); "Ecolotto" (age 7 and up), for learning about protecting endangered animals; "Save the Whales" (age 8 and up), demonstrating whales' vulnerable existence; and "Back to the Farm," about organic farming. Catalog available.

**Bongers** (P.O. Box 84366, Los Angeles, CA 90073; 213-823-0932) produces "Save the World," a cooperative board game emphasizing ecological problems (the greenhouse effect, pollution, rain forest destruction, overpopulation, etc.) and requiring teamwork to stave off ecological threats. Included is a 48-page booklet examining "eco-threats." For ages 13 and up. Price is $20 to $24; mention *The Green Consumer* and receive a 20 percent discount.

**Do Dreams Music Company** (P.O. Box 5623, Takoma Park, MD 20912; 301-445-3845) sells "Billy B." tapes with upbeat melodies and lyrics. Tapes available are "Recycle Mania" ("So much trash, so much waste, filling up the land, taking up space; so much of it could be reused but that depends on me and you; Now there's just one question we all have to ask; If we can put a man on the moon, why can't we handle our trash?")* and "Billy B. Sings About Trees," featuring lively songs like "My Roots Run Deep," "The Rock and Roll of Photosynthesis," and "Life of the Dead Tree."

*"Just One Question," Copyright 1987 Brennan and Seydewitz.

**International Video Network** (2242 Camino Ramon, San Ramon, CA 94583; 415-866-1121) has co-produced with Reader's Digest a series of "Great National Parks" videos. The hour-long "trips" through the Grand Canyon, Yosemite, and Yellowstone include park scenes from all seasons. They are available from International Video or through mail-order catalogs such as **Hammacher-Schlemmer** (800-543-3366), **Haverhills** (800-888-9920), and **Wireless** (800-669-9999).

**The Nature Company** (headquarters: P.O. Box 2310, Berkeley, CA 94702; 415-644-1337, 800-227-1114; 25 stores nationwide) offers a wide range of products, from games to books to clothing, on a variety of subjects. Write for a free catalog.

**North Carolina Biological Supply Co.** (2700 York Rd., Burlington, NC 27215; 800-334-5551, 800-632-1231 in NC) offers "Pollution," a game simulating environmental pollution and measures taken to control it. Players assume the roles of industries, businesses, and pollution-control agencies. For junior high and up. Another game is "Endangered Species," an introduction to animal conservation.

**World Wildlife Fund** (P.O. Box 224, Peru, IN 46970; 800-833-1600) sells "Down in the Rainforest," a deck of 48 game cards showing the threatened inhabitants of the rain forests. For ages 4 through 10.

## Recycled Paper Products

**Acorn Designs** (5066 Mott Evans Rd., Trumansburg, NY 14886; 607-387-3424) offers stationery, note pads, bookmarks, note cards, gift tags, and prints made with recycled paper.

**Conservatree** (10 Lombard St., Suite 250, San Francisco, CA 94111; 800, 522-9200, 415-433-1000) sells office and printing paper, including letterheads, envelopes (both plain and window), computer paper, copy paper, offset, coated papers, and newsprint, as well as text and cover papers appropriate for manuals, flyers, newsletters, and brochures. Conservatree will do an environmental impact analysis of office paper usage showing the positive environmental impacts of switching from virgin paper to recycled paper.

### *Beware of Endangered Species*

Some of the loveliest gifts are also the most harmful to any of several endangered species. Here are some gift items to avoid:

❑ **Ivory** from elephant tusks is responsible for killing between 200 and 300 African elephants per day. At that rate, the species will be extinct within twenty years, according to the World Wildlife Fund. In 1989, the United States and more than a dozen other countries declared bans on ivory imports, although a variety of ivory products still manage to find their way into American borders, and many tourists purchase carved trinkets made of ivory in Hong Kong and Japan, among the leading ivory-consuming nations. By purchasing—or even wearing—ivory, no matter how small, you are encouraging the slaughter of elephants.

In 1989, a coalition of fashion designers and retailers ran full-page ads in many newspapers proclaiming LET'S KEEP IVORY OUT OF FASHION. The list of sponsors includes:

Linda Allard for Ellen Tracy
Frederick Atkins Inc. and Atkins Member Store Corporations
Bill Blass
Dana Buchman
Carter Hawley Hale
Dayton Hudson Department Stores Co.
Louis Dell'Olio for Anne Klein & Co.

**Earth Care Paper, Inc.** (P.O. Box 3335, Madison, WI 53704; 608-256-5522), offers a variety of paper products including stationery, writing pads, office paper, journals, gift wrap, and greeting cards. Earth Care's catalog contains a wealth of information on recycling and environmental concerns; Earth Care also offers free literature on recycling and environmental organizations. Ten percent of profits is given to organizations working to solve environmental problems.

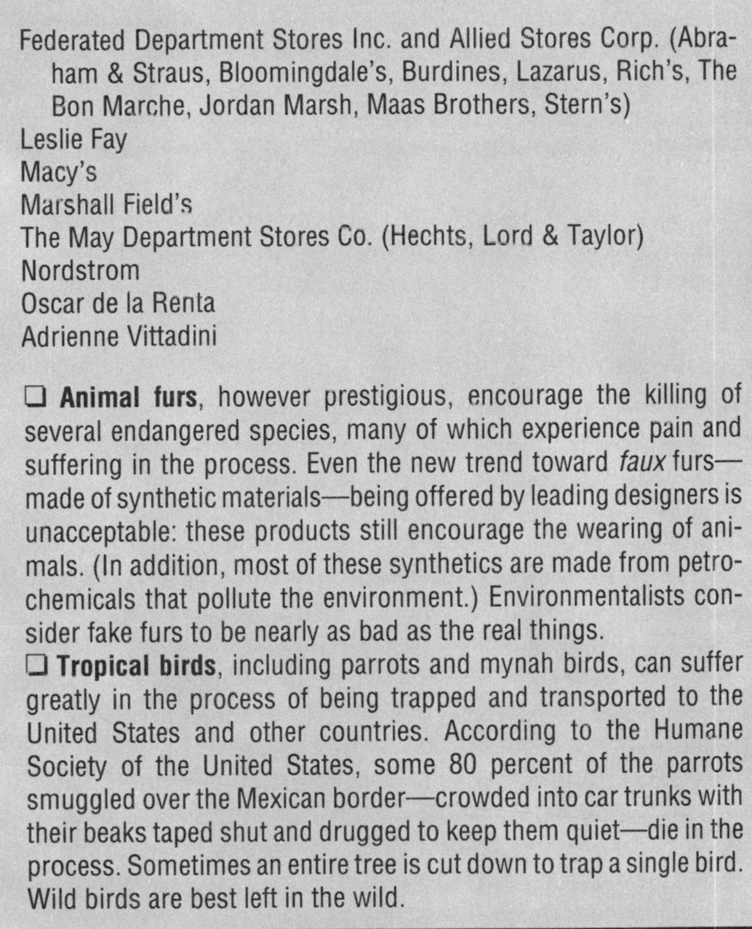

Federated Department Stores Inc. and Allied Stores Corp. (Abraham & Straus, Bloomingdale's, Burdines, Lazarus, Rich's, The Bon Marche, Jordan Marsh, Maas Brothers, Stern's)
Leslie Fay
Macy's
Marshall Field's
The May Department Stores Co. (Hechts, Lord & Taylor)
Nordstrom
Oscar de la Renta
Adrienne Vittadini

❑ **Animal furs**, however prestigious, encourage the killing of several endangered species, many of which experience pain and suffering in the process. Even the new trend toward *faux* furs—made of synthetic materials—being offered by leading designers is unacceptable: these products still encourage the wearing of animals. (In addition, most of these synthetics are made from petrochemicals that pollute the environment.) Environmentalists consider fake furs to be nearly as bad as the real things.

❑ **Tropical birds**, including parrots and mynah birds, can suffer greatly in the process of being trapped and transported to the United States and other countries. According to the Humane Society of the United States, some 80 percent of the parrots smuggled over the Mexican border—crowded into car trunks with their beaks taped shut and drugged to keep them quiet—die in the process. Sometimes an entire tree is cut down to trap a single bird. Wild birds are best left in the wild.

**Co-Op America** (10 Farrell St., South Burlington, VT 05403; 802-658-5507) distributes recycled paper products, including hand-made writing pads by the John Rossi Company; and gift wrap and greeting cards from Earth Care Paper Company. Catalog $1.

**The John Rossi Company** (259 Washburn Rd., Briarcliff Manor, NY 10510; 914-941-1752) produces "Revived Page" stationery, envelopes, hand-made writing pads and books, and paper place mats. The warm gray- and earth-toned paper is made principally from discarded newsprint. Products are available in stationery stores, museum shops, and some mail-order companies, such as **Earth Care Paper Company** and **Seventh Generation**.

**Seventh Generation** (10 Farrell St., South Burlington, VT 05403; 802-862-2999) distributes a variety of products made from recycled paper, including "low-impact stationery," personal journals, desk notes, and a stationery portfolio, all made by the John Rossi Company. Seventh Generation also sells toilet paper made from recycled paper.

## Solar-Powered Gadgets

**Chronar Sunenergy** (P.O. Box 177, Princeton, NJ 08542; 609-799-8800, 800-247-6627) produces a solar garden lamp, useful for walkways and patios. The lamp, sold in Ace, True Value, and other hardware stores, can provide five to six hours of light per night.

**Jade Mountain** (P.O. Box 4616, Boulder, CO 80306; 303-449-6601) offers various solar-powered gifts, including a speedboat, a "cool cap," a safari hat, a UFO hot-air balloon, a tea jar, flashlights, battery chargers, and a "4-in-1 Solar Construction Kit," for children to build solar-powered windmills, airplanes, and helicopters.

**Solar Electric Engineering** (175 Cascade Ct., Rohnert Park, CA 94928; 707-586-1987, 800-832-1986) sells a solar-powered sports radio with earphones. The radio requires three hours in the sun to generate enough power for four hours of playing time; it also runs on conventional AA batteries.

**SunWatt Corporation** (RFD Box 751, Addison, ME 04606; 207-497-2204) offers solar battery chargers and rechargeable batteries.

## Unusual Gifts

**The Basic Foundation** (P.O. Box 47012, Saint Petersburg, FL 33743; 813-526-9562) plants hardwood and fruit trees in endangered forests in Costa Rica (see "Garden and Pet Supplies" chapter for more information). A gift of $5 pays for a single tree; $250 for planting a hectare—containing approximately 1,000 trees.

**Programme for Belize** (P.O. Box 1088, Vineyard Haven, MA 02568; 508-693-0856) offers a chance to buy an acre of tropical rain forest. For each $50 contribution, a certificate will be sent stating that an acre (or more) has been protected. The acre will be part of a 110,000-acre tract of forest in Belize, which will be developed into a park and model of sustained yield development.

**The Whale Center** (3933 Piedmont Ave., Suite 2, Oakland, CA 94611; 415-654-6621), a nonprofit organization supporting whales, has an Adopt-a-Gray-Whale program. For a $24 "adoption," donors receive a color photograph of the "adopted" whale, an adoption certificate, and information about the whale, including quarterly updates and activity sheets. Proceeds directly benefit the center.

## Gifts That Support Environmental Organizations

Many environmental and wilderness conservation organizations offer gift catalogs, with products ranging from T-shirts and sweatshirts to posters, slides, books, calendars, films and videos, tote bags, mugs, umbrellas, and children's items. Proceeds from sales of these items usually go toward the organization's operating and project costs. Contact these groups for their current catalogs.

Center for Marine Conservation
Whale Gifts Catalog Dept.
P.O. Box 810
Old Saybrook, CT 06575
203-388-4436

The Cousteau Society
930 W. 21st St.
Norfolk, VA 23517
804-627-1144

Environmental Action
1525 New Hampshire Ave. NW
Washington, DC 20036
202-745-4870

Greenpeace
P.O. Box 6012
San Francisco, CA 94101
415-474-1870, 800-323-1776

National Audubon Society
Expedition Institute
Sharon, CT 06069
203-364-0522

The Nature Conservancy
1815 N. Lynn St. Dept. M-4
Arlington, VA 22209
703-841-8747

Rainforest Action Network
301 Broadway, Suite A
San Francisco, CA 94133
415-398-4404

Save Our Ecosystems, Inc.
541 Willamette St., #315
Eugene, OR 97401
503-484-2679

Sierra Club
Public Affairs
730 Polk St.
San Francisco, CA 94109
415-776-2211

World Wildlife Fund
P.O. Box 224
Peru, IN 46970
800-833-1600

# H·O·M·E E·N·E·R·G·Y AND F·U·R·N·I·S·H·I·N·G·S

There is no place like home when it comes to being a Green Consumer. Aside from the food we bring into it, the trash we take out of it, the gardens we plant around (or inside) it, and the vehicles in which we drive to and from it, there is the matter of the home itself—in particular, the energy needed to keep its occupants' lives cozy and comfortable.

The cost of that energy is not insignificant, by any measure. Homes are the third largest users of energy (after industry and transportation), representing about 20 percent of all energy consumed in the United States, at a cost of about $130 billion, just under 3 percent of our nation's gross national product. Producing that energy—through the thousands of coal-fired, nuclear, hydroelectric, and other power-generating facilities—costs the earth dearly in terms of the resources it requires and the pollution it creates. Electrical utilities are primary contributors to acid rain (through the sulfur dioxide emissions of coal-fired power plants) and greenhouse warming (through the burning of fossil fuels such as coal, oil, and natural gas). (See "Seven Environmental Problems You Can Do Something About" for more on this.)

Worst of all, such problems needn't exist. Estimates of the amount of energy wasted in U.S. homes through inefficiency range from 40 to 70 percent. Conservatively, most experts agree that we could easily save about half of our home energy costs—along with similar cuts in pollutants—if we used our energy resources wisely, including purchasing energy-efficient appliances. That wasted energy is also wasted money—hundreds of dollars a year for some households. As you'll see, the purchase price of most appliances pales compared to the cost of operating them over their lifetimes.

Wasted energy isn't the only item about which Green Consumers should be concerned at home. Water use is another big concern, particularly in drought-prone sections of the United States. Those areas used to be limited to western states, but in recent years the Midwest and the populous portions of the East Coast also have faced water-shortage problems, with most forecasts indicating even greater problems still to come.

There's more. Many home furnishings, from carpets to couches to cabinets, are made from materials that create additional pollutants either in their manufacturing process or when disposed of, or which come from threatened rain forest timber. If you understand the potential problems, it is not difficult to choose products that don't contain these environmentally unfriendly components.

In short, there's plenty that Green Consumers can do right in their own homes, with little inconvenience or disruption to their lifestyles.

## Energy-Efficient Appliances

Equipping your home with electrical and gas appliances that use less energy is more than environmentally sound—it's downright profitable. Appliances and heating and cooling equipment cost an average American household more than $1,000 per year. You can sharply reduce those costs—and help reduce the need for building polluting and resource-draining power plants by using high-efficiency appliances.

You may be surprised to learn which appliances in your home guzzle the most electricity. You may have been taught always to turn off lights when you leave the room, for example, but lights are the least of the problem. Your refrigerator, quietly (or not-so-quietly) gurgling in the corner of the kitchen has a far bigger electrical appetite. In fact, cumulatively, America's refrigerators use the output of about 25 large power plants—about 7 percent of the nation's total electricity consumption and more than half of the power generated by the nation's nuclear power plants. According to one estimate, if every household in the United States had the most energy-efficient refrigerators currently on the market, the electricity savings would eliminate the need for about ten large power plants.

Beginning in 1990, federal law establishes minimum efficiency standards for major home appliances and heating and cooling equipment. (The standards for air conditioners and furnaces are scheduled to take place in 1992.) The standards will require all new appliances to be 10 to 30 percent more efficient than the average models sold previously. It has been estimatedthat between now and the year 2000, these standards will save consumers at least $28 billion in energy costs over the lifetimes of the products—about $300 per household. Moreover, the federal standards will reduce peak electricity demand by the equivalent of 25 large power plants.

There's no question that energy-efficient appliances cost more to purchase than inefficient ones, but that scarcely matters: *the purchase price is by far the smallest cost of owning an appliance.* For example, the cost of electricity to run a refrigerator for 15 to 20 years typically costs three times as much as the refrigerator's purchase price. (See the box "Determining Life Cycle Costs.") So buying efficient appliances is a good long-term investment. And as energy costs continue to rise—the average residential electricity rate has more than tripled during the past 15 years and is certain to continue going up—the investment in energy-efficient appliances will yield even bigger dividends.

**The Ten Key Energy Users.** Not including your central heating system, here are the ten things likely to be the biggest energy users in your home:

1. Water heaters
2. Refrigerator/freezers
3. Freezers
4. Air conditioners
5. Ranges
6. Clothes washers
7. Clothes dryers
8. Dishwashers
9. Portable space heaters
10. Lighting

## Comparing Appliances

How can you tell which appliances are the most efficient? Seven major appliances—boilers, clothes washers, dishwashers, heat pumps, refrigerators and refrigerator/freezers, room air conditioners, and water heaters—are required by law to disclose their energy efficiency ratings (EERs) or annual energy costs on yellow Energy Guide labels attached to new appliances. (See sample label on page 178.) These costs and ratings are derived from standardized industry tests conducted by manufacturers. Other appliances are more difficult to compare, although there are energy-saving features worth looking for.

Beginning on page 179 are listings of the most energy-efficient appliances in ten categories: refrigerators, freezers, clothes washers, dishwashers, gas furnaces, oil furnaces, water heaters, room air conditioners, central air conditioners, and central heat pumps. Due to space limitations, only the five top-rated models (in terms of energy efficiency) in each category are included. These models represent fewer than 5 percent of all models currently available. The information was selected from a much larger list, compiled by the **American Council for an Energy-Efficient Economy** (1001 Connecticut Ave. NW, Suite 535, Washington, DC 20036; 202-429-8873) and is reprinted here with permission from ACEEE. To obtain the most recent complete list of appliance energy ratings, send $2 to ACEEE for its booklet *The Most Energy-Efficient Appliances*. Another helpful ACEEE booklet is *Saving Energy and Money With Home Appliances*, which is $3.

Keep in mind that energy performance is only one of several important criteria for selecting home appliances. These ratings, for example, do not consider product reliability and other features such as freezer capacity, access to shelves, and options such as ice makers and specialized storage bins. However, energy-efficient appliances are by nature high-quality products due to the advanced materials and manufacturing processes used in their construction. To compare other appliance features, you might consider consulting *Consumer Reports*, which regularly provides in-depth analyses of major appliances.

An asterisk (*) appearing in a model number indicates a digit or letter that varies with the color or style of the appliance.

## *Determining Life-Cycle Costs*

The real cost of an appliance can't be judged when it is purchased, or even over the course of a single year. The most accurate measure is the "life-cycle cost"—the cost to own and operate an appliance over its lifetime. In short, the life cycle cost is:

**Purchase Price + Annual Energy Cost X Estimated Lifetime = Life-cycle Cost**

For example, if you purchase a refrigerator for $650, it costs about $150 a year to run. If it lasts for 20 years, the life-cycle cost would be

$650 + ($150 X 20) = $3,650

To help you determine life-cycle costs, here are the average life spans of major appliances:

| Appliance | Average life span (years) |
|---|---|
| Air conditioner (central) | 15 |
| Air conditioner (room) | 12 |
| Clothes dryer | 18 |
| Clothes washer | 13 |
| Dishwasher | 12 |
| Freezer | 20 |
| Range/oven | 18 |
| Refrigerator/freezer | 20 |
| Water heater | 13 |

Compare, for example, Refrigerator A, which costs $500 and costs $125 a year in energy, with Refrigerator B, which costs $750 and costs $105 a year in energy. Life-cycle costs would be as follows:

*Refrigerator A = $500 + ($125 X 20) = $3,000*
*Refrigerator B = $750 + ($105 X 20) = $2,850*

So, even though Refrigerator B costs $250 more to buy, the life-cycle costs will actually be $150 less than those for Refrigerator A over the life cycle of the appliance.

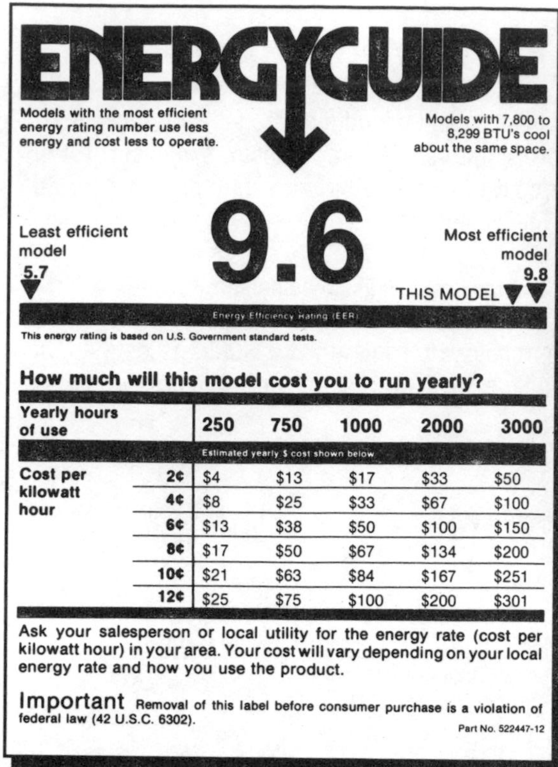

ENERGYGUIDE

Models with the most efficient energy rating number use less energy and cost less to operate.

Models with 7,800 to 8,299 BTU's cool about the same space.

Least efficient model 5.7

9.6

Most efficient model 9.8

THIS MODEL

Energy Efficiency Rating (EER)

This energy rating is based on U.S. Government standard tests.

**How much will this model cost you to run yearly?**

| Yearly hours of use | | 250 | 750 | 1000 | 2000 | 3000 |
|---|---|---|---|---|---|---|
| | | Estimated yearly $ cost shown below | | | | |
| Cost per kilowatt hour | 2¢ | $4 | $13 | $17 | $33 | $50 |
| | 4¢ | $8 | $25 | $33 | $67 | $100 |
| | 6¢ | $13 | $38 | $50 | $100 | $150 |
| | 8¢ | $17 | $50 | $67 | $134 | $200 |
| | 10¢ | $21 | $63 | $84 | $167 | $251 |
| | 12¢ | $25 | $75 | $100 | $200 | $301 |

Ask your salesperson or local utility for the energy rate (cost per kilowatt hour) in your area. Your cost will vary depending on your local energy rate and how you use the product.

**Important** Removal of this label before consumer purchase is a violation of federal law (42 U.S.C. 6302).

Part No. 522447-12

## Refrigerators

Today's energy-efficient refrigerators use about one-third less electricity than comparable models ten years old or older. To gain the maximum efficiency from a refrigerator, it's important to shop for the size and features that best suit your needs. If you buy one that's too big, you'll waste energy. A refrigerator that's too small won't cool foods as efficiently. A rule of thumb is that one to two people need a model with an interior capacity of at least 12 cubic feet. Three to four people need 14 to 16 cubic feet. For more than four people, add two cubic feet of space for each additional person.

The models listed below are grouped by door style, defrosting capability, and volume (size). Models are listed within each group in order of increasing electricity use. If two different-size refrigerators use the same amount of electricity per year, the larger model would be considered more efficient, because it keeps more space cold with the same amount of energy.

The annual energy costs shown below are based on an electricity price of 6.75 cents per kilowatt/hour (kwh), which is the electricity price used on the yellow Energy Guide labels. Your actual energy cost may differ depending on the price of electricity in your area and your use of the refrigerator.

## *Single Door, Manual Defrost 10.4–14.4 Cubic Feet*

| Brand | Model | Cu. Ft. | Kwh/Yr | Annual Energy Cost |
|---|---|---|---|---|
| Kenmore | 86111*2 | 11.0 | 367 | $28 |
| Kenmore | 86311*0 | 11.0 | 434 | 33 |
| U-Line | UR-110 | 11.6 | 434 | 33 |
| Kenmore | 86611*0 | 11.6 | 489 | 38 |
| Sanyo | SR 1057 | 11.6 | 489 | 38 |

## *Top Freezer, Partial Automatic Defrost 12.5–16.4 Cubic Feet*

| Brand | Model | Cu. Ft. | Kwh/Yr | Annual Energy Cost |
|---|---|---|---|---|
| Kenmore | 83831* | 13.7 | 735 | $57 |
| General Electric | TB13S* | 13.4 | 735 | 57 |
| Hotpoint | CTA13C* | 13.4 | 735 | 57 |
| Frigidaire | FCD-14TF*_* | 14.0 | 766 | 59 |
| White-Westinghouse | RT142G**5 | 14.0 | 767 | 59 |
| General Electric | TB15S* | 14.6 | 770 | 59 |

## *Top Freezer, Automatic Defrost 14.5–16.4 Cubic Feet*

| Brand | Model | Cu. Ft. | Kwh/Yr | Annual Energy Cost |
|---|---|---|---|---|
| Frigidaire | FPI-16TE*-0 | 16.0 | 766 | $59 |
| Kelvinator | TPK160BN** | 16.0 | 766 | 59 |
| Kenmore | 86962** | 16.0 | 766 | 59 |
| Kenmore | 87960*0 | 16.0 | 766 | 59 |
| Kenmore | 87662** | 16.0 | 766 | 59 |

Annual energy costs based on fuel rate of 7 cents per kilowatt-hour. Asterisk indicates a number or letter indicating model color or other nonessential feature.

Source: American Council for an Energy-Efficient Economy, 1989

## Tips for Refrigerator Efficiency

❑ **Keep the condenser coils clean.** (The coils are on the back or at the bottom of the refrigerator.) At least once a year, carefully wipe, vacuum, or brush the coils to remove dust and dirt. Many refrigerators provide access to coils through a removable panel.

❑ **Make sure the door gasket is clean and tight.** This will ensure that no cold air is escaping into the room from the refrigerator. To check the tightness, close the refrigerator door with a single sheet of paper placed between the door and the cabinet. With the door closed, try pulling on the paper. It should come loose with a bit of effort. If it comes loose too easily, you may need to replace the gasket or adjust the hinges. Do this test in several places along the door.

❑ **Check the temperature.** You may wish to keep a thermometer in the refrigerator to check the temperature periodically. The ideal temperature for the refrigerator is between 38 degrees and 40 degrees Fahrenheit; the freezer ideally should be 0 degrees Fahrenheit.

## 16.5–18.4 Cubic Feet

| Brand | Model | Cu. Ft. | Kwh/Yr | Annual Energy Cost |
|---|---|---|---|---|
| Gibson | RT17F**U3A | 16.6 | 766 | $59 |
| Frigidaire | FPD-17TIF-0 | 16.6 | 766 | 59 |
| Frigidaire | FP*-17TF*-0 | 16.6 | 766 | 59 |
| White-Westinghouse | RT1742C*1 | 16.6 | 766 | 59 |
| Frigidaire | FPCI-18TTE*-1 | 18.0 | 840 | 65 |

## 18.5–20.4 Cubic Feet

| Brand | Model | Cu. Ft. | Kwh/Yr | Annual Energy Cost |
|---|---|---|---|---|
| Frigidaire | FP*-19TF*-0 | 18.6 | 839 | $65 |
| Frigidaire | FPC*-19TF*-0 | 18.6 | 839 | 65 |
| Frigidaire | FPES-19TF*-0 | 18.6 | 839 | 65 |
| Frigidaire | FPCI-19TIF*-0 | 18.6 | 839 | 65 |
| Frigidaire | FPE-19TIF*-0 | 18.6 | 839 | 65 |

## *20.5–24.4 Cubic Feet*

| Brand | Model | Cu. Ft. | Kwh/Yr | Annual Energy Cost |
|---|---|---|---|---|
| General Electric | TBX21*KC | 20.7 | 941 | $73 |
| Hotpoint | CTX21*KC | 20.7 | 941 | 73 |
| Kenmore | 86*15* | 20.6 | 942 | 73 |
| General Electric | TBX22ZKC | 21.7 | 944 | 73 |
| Hotpoint | CTX22GKC | 21.7 | 944 | 73 |

## *Side-By-Side, Automatic Defrost 18.5–20.4 Cubic Feet*

| Brand | Model | Cu. Ft. | Kwh/Yr | Annual Energy Cost |
|---|---|---|---|---|
| Amana | S*I20J | 19.9 | 1032 | $80 |
| Amana | SCTI20H | 19.9 | 1032 | 80 |
| Amana | SBI20H | 19.9 | 1032 | 80 |
| Amana | SC19* | 19.4 | 1100 | 85 |
| Frigidaire | FPZ-19VF*-0 | 19.0 | 1127 | 87 |

## *20.5–22.4 Cubic Feet*

| Brand | Model | Cu. Ft. | Kwh/Yr | Annual Energy Cost |
|---|---|---|---|---|
| Maytag | RSD22A | 21.8 | 1117 | $86 |
| Holland Dist. | NDNT2292 | 22.0 | 1117 | 86 |
| Maytag | RSW22A | 21.6 | 1147 | 88 |
| Amana | S*22* | 22.4 | 1159 | 89 |
| Amana | SL22JB | 22.4 | 1159 | 89 |

### 22.5–26.4 Cubic Feet

| Brand | Model | Cu. Ft. | Kwh/Yr | Annual Energy Cost |
|---|---|---|---|---|
| Frigidaire | FPCE-24VF*-0 | 24.0 | 1117 | $86 |
| Maytag | RSD24A | 23.8 | 1123 | 86 |
| Amana | SC25JP | 25.2 | 1199 | 92 |
| Amana | S*25* | 25.2 | 1199 | 92 |
| Amana | S**25H | 25.2 | 1199 | 92 |

Annual energy costs based on fuel rate of 7 cents per kilowatt-hour. Asterisk indicates a number or letter indicating model color or other nonessential feature.

## Freezers

The energy efficiency of freezers has improved over the past decade, although at a slower rate than refrigerators.

### Upright, Manual Defrost 13.5–17.4 Cubic Feet

| Brand | Model | Cu. Ft. | Kwh/Yr | Annual Energy Cost |
|---|---|---|---|---|
| Frigidaire | UFS16D-W*3 | 15.8 | 585 | $45 |
| Frigidaire | UFE16D-L** | 16.1 | 610 | 47 |
| Montgomery Ward | FFT46858-* | 16.1 | 610 | 47 |
| White-Westinghouse | FU168*R** | 16.1 | 610 | 47 |
| Amana | ESU15D | 15.0 | 649 | 50 |

### 17.5–21.4 Cubic Feet

| Brand | Model | Cu. Ft. | Kwh/Yr | Annual Energy Cost |
|---|---|---|---|---|
| Kenmore | 82639** | 19.2 | 725 | $56 |
| Whirlpool | EV190E*S*O* | 19.2 | 725 | 56 |
| Whirlpool | EV190F*S*O* | 19.2 | 725 | 56 |
| Wood's | *U60 | 20.9 | 796 | 61 |
| Home Products Co. | HPV21-A | 20.9 | 796 | 61 |

## *Safe Disposal of Refrigerators*

What do you do with your old worn-out refrigerator? Granted, it's not a problem people face many times in their lives—refrigerators can last for decades—but when the time does come, how you dispose of it can have a major impact on the environment. Refrigerators (as well as freezers and air conditioners) contain chlorofluorocarbons, and when the appliances are thrown into a trash heap or otherwise discarded, CFCs can be released into the atmosphere. All told, about 3 million pounds of CFCs are released into the atmosphere by the approximately 6 million refrigerators tossed out each year by Americans. Moreover, capacitors in some refrigerators manufactured before 1980 contain small amounts of polychlorinated biphenyls, or PCBs, which have been classified as a hazardous substance by the Environmental Protection Agency.

**Appliance Recycling Centers of America** (ARCA, 654 University Ave. West, Saint Paul, MN 55104; 612-291-1100) is finding a profitable way to save CFCs and other substances from the environment, as well as ensuring that countless tons of scrap metal are recycled and reused. ARCA—with centers in Atlanta, Cleveland, Milwaukee, Saint Louis, and Saint Paul—picks up refrigerators and other appliances at curbside and removes CFC-containing Freon as well as PCB-laden capacitors, insulation, copper parts, and mercury-containing electrical switches. The hazardous wastes are put into approved containers and hauled to a hazardous-waste collection site. The remaining scrap is then recycled as scrap. About 10 percent of the appliances ARCA picks up are conditioned for resale.

## *Upright, Automatic Defrost 13.5–17.4 Cubic Feet*

| Brand | Model | Cu. Ft. | Kwh/Yr | Annual Energy Cost |
|---|---|---|---|---|
| Amana | ESUF16D | 16.2 | 871 | $67 |
| Kitchen Aid | K*FF15MS**Y* | 14.9 | 918 | 71 |
| Frigidaire | UFP16D-L** | 15.7 | 934 | 72 |
| General Electric | CAF16D* | 15.7 | 934 | 72 |
| Gibson | FV16F5**F* | 15.7 | 934 | 72 |

## Chest, Manual Defrost
## 13.5–17.4 Cubic Feet

| Brand | Model | Cu. Ft. | Kwh/Yr | Annual Energy Cost |
|---|---|---|---|---|
| Home Products | HPC 16A | 16.8 | 428 | $33 |
| Panasonic | NR-1705 FC | 16.8 | 430 | 33 |
| Whirlpool | EH150FXVN00 | 15.0 | 454 | 35 |
| Wood's | *C42 | 15.1 | 459 | 35 |
| Whirlpool | EH150F*V*O* | 15.1 | 459 | 35 |

## 17.5–23.4 Cubic Feet

| Brand | Model | Cu. Ft. | Kwh/Yr | Annual Energy Cost |
|---|---|---|---|---|
| Panasonic | NR-2105FC | 20.7 | 519 | $40 |
| Home Products | HPC 20A | 20.7 | 521 | 40 |
| Wood's | OC58 | 20.7 | 521 | 40 |
| Home Products | HPC 20 | 20.3 | 529 | 41 |
| Whirlpool | EH220F*V*O* | 22.1 | 579 | 45 |

Annual energy costs based on fuel rate of 7 cents per kilowatt-hour. Asterisk indicates a number or letter indicating model color or other nonessential feature.

## Dishwashers

Most of the cost of running dishwashers (and clothes washers) is for heating water. So, an efficient water heater is essential to reducing energy use. The annual energy costs shown are based on a typical electric water heater and an electricity price of 7.63 cents per kilowatt-hour, the price used on the most recent Energy Guide labels for dishwashers.

## *Dishwashers*

| Brand | Model | Kwh/Yr. | Annual Energy Cost |
|---|---|---|---|
| White-Westinghouse | SU200JX*2 | 501 | $38 |
| Caloric | DUS314-19 | 639 | 49 |
| Frigidaire | DW180***1 | 639 | 49 |
| Admiral | DU118J4 | 639 | 49 |
| Crosley | DU18J4 | 639 | 49 |

Annual energy costs based on fuel rate of 7 cents per kilowatt-hour. Asterisk indicates a number or letter indicating model color or other nonessential feature.

## Clothes Washers

As with dishwashers, the biggest cost is hot water. The annual energy costs shown are based on a typical electric water heater and an electricity price of 7.63 cents per kilowatt-hour, the price used on the most recent Energy Guide labels for clothes washers.

## *Compact Size*
**(less than 16-gallon capacity)**

| Brand | Model | Kwh/Yr. | Annual Energy Cost |
|---|---|---|---|
| General Electric | 213JB 118 | 623 | $50 |
| Gibson | 213JB 24F | 623 | 50 |
| Kelvinator | 213JB C340 | 623 | 50 |
| Kelvinator | 213JB AW12Q1 | 623 | 50 |
| Montgomery Ward | 213JB 6507 | 623 | 50 |

Annual energy costs based on fuel rate of 7 cents per kilowatt-hour. Asterisk indicates a number or letter indicating model color or other nonessential feature.

## Standard Size
### (16 or more gallons capacity)
### *Front Loading*

| Brand | Model | Kwh/Yr. | Annual Energy Cost |
|---|---|---|---|
| Gibson | WS 27 M6-V | 291 | $23 |
| White-Westinghouse | LT*OOL | 291 | 23 |
| White-Westinghouse | LT250L | 291 | 23 |
| White-Westinghouse | LT150L | 299 | 24 |

### *Top Loading*

| Brand | Model | Kwh/Yr. | Annual Energy Cost |
|---|---|---|---|
| Montgomery Ward | LNC6105A | 651 | $52 |
| Maytag | A212S | 720 | 58 |
| Maytag | A190 | 820 | 66 |
| Maytag | A312 | 820 | 66 |
| Maytag | A21* | 820 | 66 |

Annual energy costs based on fuel rate of 7 cents per kilowatt-hour. Asterisk indicates a number or letter indicating model color or other nonessential feature.

Source: American Council for an Energy-Efficient Economy, 1989

## Water Heaters

In recent years, new types of water heaters have come onto the market. In addition to the traditional storage water heater—in which a ready reserve of water is always kept hot—a new type of "demand" water heater heats water only when it is needed. While hot water never runs out, the rate of flow is limited. This trade-off results in an energy savings of about 25 percent, or about $75 per year for gas appliances, a bit more for electrical appliances.

There are other energy-saving water heater technologies worth considering. Some newer systems are designed to take advantage of a home's primary heating system to heat hot water. Solar water heaters use the sun's heat to produce hot water, often in combination with a backup conventional heating unit during cold periods with little sunshine. Heat-pump water heaters take heat from the air and move it to the water in a storage tank. Each of these has advantages (they are energy efficient and less expensive than traditional water heaters) and disadvantages (they don't always get water as hot as traditional water heaters) and are worth considering if you plan to install a new water heater.

The energy efficiency of a water heater is indicated by its Energy Factor, or "EF," an overall number based on the use of 64 gallons of hot water per day. The average new gas water heater sold in 1988 had an EF of about 0.85. Under the new federal standards, the minimum EF ratings for today's water heaters are based on storage tank size. In general, the smaller the tank size, the higher the EF. (Higher numbers indicate better efficiency.)

### *Minimum Energy Factors for Water Heaters*

| Tank Size | Gas | Oil | Electric |
|---|---|---|---|
| 30 gallons | 0.56 | 0.53 | 0.91 |
| 40 gallons | 0.54 | 0.53 | 0.90 |
| 50 gallons | 0.53 | 0.50 | 0.88 |
| 60 gallons | 0.51 | 0.48 | 0.87 |

Choosing the right tank size is very important. If it is too large, you will waste a lot of energy heating water you'll never use. If it's too small, you're bound to run out of hot water in the middle of your shower. The ability of hot water heaters to meet peak demands for hot water is indicated by the "first hour rating" (listed in the "rating" column below). This rating accounts for the effects of tank size and the speed by which cold water is heated. In some cases a water heater with a small tank but a powerful burner or heating element can have a higher first-hour rating than a unit with a large tank and a less-powerful burner or heating element. (For more information on first hour ratings, contact the **Gas Appliance Manufacturers Association**, 1901 N. Moore St., Arlington, VA 22209; 703-525-9565.) The models listed below are grouped by their tank size.

## Gas Water Heaters

| Brand | Model | Rating (Gal./Hr.) | Tank Size (Gallons) | Efficiency (EF) |
|---|---|---|---|---|
| Bradford-White | M-III-303T5CN-7 | 64 | 29 | 0.64 |
| Rheem | 21VR30-5 | 52 | 30 | 0.64 |
| Rheem | 21VR30-5N | 52 | 30 | 0.64 |
| U.S. Water Heater | M-III-303T5CN-7 | 64 | 29 | 0.64 |
| American Appliance | GXN363* | 55 | 30 | 0.63 |

## Oil Water Heaters

| Brand | Model | Rating (Gal./Hr.) | Tank Size (Gallons) | Efficiency (EF) |
|---|---|---|---|---|
| Bock | 32PP | 131 | 32 | 0.63 |
| Bradford-White | F-I-30SE50 | 120 | 30 | 0.62 |
| Carlin | RCG-30 | 120 | 30 | 0.62 |
| Ford | FC3016E | 120 | 30 | 0.62 |
| Rheem | OGL-30 | 120 | 30 | 0.62 |

## Electric-Resistance Water Heaters

| Brand | Model | Rating (Gal./Hr.) | Tank Size (Gallons) | Efficiency (EF) |
|---|---|---|---|---|
| Marathon | M30238 | 39 | 30 | 0.98 |
| Marathon | MP302*5 | 42 | 30 | 0.98 |
| Rheem | 81S*30DT | 35 | 30 | 0.97 |
| Rheem | 81G*30* | 41 | 30 | 0.97 |
| Rheem | H1VG30-* | 41 | 30 | 0.97 |

## Heat-Pump Water Heaters

| Brand | Model | Rating (Gal./Hr.) | Tank Size (Gallons) | Efficiency (EF) |
|---|---|---|---|---|
| Therma-Stor (DEC) | TS-HP-52-HR | 48 | 52 | 3.2 |
| Reliance | 5521AHP41HP3 | 54 | 52 | 3.0 |
| State | SS8521AHP4CHP | 54 | 52 | 3.0 |
| Therma-Stor (DEC) | TS-HP-80-HRA | 74 | 80 | 3.5 |
| Therma-Stor (DEC) | TS-HP-120-18 | 103 | 120 | 3.5 |

Asterisk indicates a number or letter indicating model color or other nonessential feature.

Source: American Council for an Energy-Efficient Economy, 1989

## Room Air Conditioners

An air conditioner takes heat from a room, using a special heat-absorbing fluid called a refrigerant (usually containing Freon or another type of chlorofluorocarbon), and moves the heat outside. As heat is removed from the room, the room becomes cooler.

In choosing a room air conditioner, you should consider two primary factors: the cooling capacity and the operating efficiency. The cooling capacity is expressed in BTU/hour; your appliance dealer should provide a Cooling Load Estimate Form from the Association of Home Appliance Manufacturers showing the correct cooling capacity for the room you plan to cool.

The operating efficiency of room air conditioners is rated by their Energy Efficiency Ratio, or EER, the ratio of the cooling output divided by the power consumption. The higher the EER, the more efficient the air conditioner. An average new room air conditioner has an EER of about 8. A model with an EER greater than 8.5 is considered "efficient"; a model with an EER over 9.5 is "very efficient." The new national standards set the minimum average EER at about 8.6.

### *Room Air Conditioners*

#### *5,000–6,999 BTU/HR Cooling Capacity*

| Brand | Model | BTU/HR | Efficiency (EER) |
|---|---|---|---|
| Sharp | AF-607M6 | 6,300 | 9.8 |
| Teknika | AK61E | 5,900 | 9.6 |
| Crosley | CCA 6E62 | 6,300 | 9.5 |
| Emerson Quiet Cool | 6CC53 | 6,300 | 9.5 |
| Montgomery Ward | VEE 5463 | 6,300 | 9.5 |

#### *7,000–8,999 BTU/HR Cooling Capacity*

| Brand | Model | BTU/HR | Efficiency (EER) |
|---|---|---|---|
| Friedrich | SS07H10A | 7,200 | 11.0 |
| Friedrich | SS08H10A | 8,200 | 10.5 |
| Kenmore | 106.8770881 | 8,000 | 9.7 |
| Teknika | AK83E | 8,000 | 9.7 |
| Kenmore | 253.878089* | 8,000 | 9.6 |

## *Maintaining Air Conditioners and Heat Pumps*

Air conditioners and heat pumps need regular maintenance to function at peak efficiency:

❑ **Room air conditioners.** Inspect the filter monthly and change or clean it. The condenser should be cleaned professionally every few years. During the winter, make sure the air conditioner or its opening is sealed tightly to keep out winter drafts.

❑ **Central air conditioners and heat pumps.** They should be inspected, cleaned, and tuned every three years by a professional service person. Besides extending the life of the system, regular servicing can lead to 10 to 20 percent energy savings. The power to a CAC should be turned off during the winter; otherwise, the heating elements will consume power all winter long. To turn the power off, flip the circuit breaker by the outside unit of the air conditioner. The power should be turned on at least one day before the air conditioner is actually used, to prevent compressor damage.

## *9,000–10,999 BTU/HR Cooling Capacity*

| Brand | Model | BTU/HR | Efficiency (EER) |
|---|---|---|---|
| Carrier | 51GM*0091 | 9,000 | 12.0 |
| Friedrich | SM10H10A | 10,300 | 12.0 |
| General Electric | ACD09LA | 9,000 | 12.0 |
| Hotpoint | KCD09LA | 9,000 | 12.0 |
| General Electric | AVX10FA | 10,000 | 11.6 |

## *11,000–12,999 BTU/HR Cooling Capacity*

| Brand | Model | BTU/HR | Efficiency (EER) |
|---|---|---|---|
| Daikin | R30ASAN/FT303ASAN | 12,500 | 10.0 |
| Friedrich | SS12H10A | 12,000 | 10.0 |
| Airtemp | LA1263VES | 12,400 | 9.7 |
| Teco | LA126*VS | 12,400 | 9.7 |
| Teco | LA126*VHS | 12,400 | 9.7 |

## *13,000–14,999 BTU/HR Cooling Capacity*

| Brand | Model | BTU/HR | Efficiency (EER) |
|---|---|---|---|
| Friedrich | SM14H10A | 14,000 | 10.5 |
| Carrier | 51GMA 1141 | 13,500 | 10.2 |
| General Electric | ACD14AA | 13,500 | 10.2 |
| Kenmore | 106.8761492 | 14,000 | 10.2 |
| Airtemp | *SL 14F2J | 14,000 | 10.0 |

## *15,000–16,999 BTU/HR Cooling Capacity*

| Brand | Model | BTU/HR | Efficiency (EER) |
|---|---|---|---|
| Carrier | 51CMA1153 | 15,000 | 9.2 |
| Comfort-Aire | WV163HE | 16,400 | 9.0 |
| Friedrich | *S15H3*A | 15,000 | 9.0 |
| General Electric | AC*15DA | 15,000 | 9.0 |
| Hotpoint | KCS15DA | 15,000 | 9.0 |

## *17,000–19,999 BTU/HR Cooling Capacity*

| Brand | Model | BTU/HR | Efficiency (EER) |
|---|---|---|---|
| Friedrich | SL19H30 | 19,000 | 9.5 |
| Panasonic | CW-1802QU | 18,000 | 9.5 |
| Quasar | HQ2182DW | 18,000 | 9.5 |
| Friedrich | EL19H35 | 18,800 | 9.4 |
| Sharp | AF-1807M8 | 18,500 | 9.3 |

Asterisk indicates a number or letter relating to model color or other nonessential feature.

Source: American Council for an Energy-Efficient Economy, 1989

## Central Air Conditioners

Central air-conditioning (CAC) units operate similarly to room air conditioners, but they are rated differently. The standard is the Seasonal Energy Efficiency Rating, or SEER, which is the cooling output divided by the power input for an average U.S. climate. The average CAC sold in 1988 had a SEER of about 9; many older units have a SEER of 6 or 7. The 1992 national efficiency standards call for a minimum SEER of 10. In general, a SEER of 10 is good; a SEER of 12 is considered to be excellent.

If you have an older, inefficient unit, do not replace its compressor unit with a newer, more efficient one unless it is compatible with the CAC's blower coil. A highly efficient compressor may not achieve its rated efficiency if paired with an older blower coil. Ask your appliance dealer or service technician for guidance.

The air conditioners below are grouped by their cooling capacity. Each "ton" represents 12,000 BTU/hour of cooling capacity.

### *Cooling Capacity: approx. 2 tons*

| Brand | Outdoor Unit | Indoor Unit | Capacity BTU/HR | Efficiency (SEER) |
|---|---|---|---|---|
| Trane | TTX724A | TW*739E15+BAY24X045 | 26,400 | 13.45 |
| Trane | TTX724A | TW*739E15-C | 26,400 | 13.45 |
| Rheem | RAJA-024JA | RHQU-08+RCTB-A024 | 23,600 | 12.85 |
| Ruud | UAJA-024JA | UHQA-08+RCTB-A024 | 23,600 | 12.85 |
| Trane | TTX724A | TWH739P15+BAY24X045 | 26,400 | 12.80 |

### *Cooling Capacity: approx. 2.5 tons*

| Brand | Outdoor Unit | Indoor Unit | Capacity BTU/HR | Efficiency (SEER) |
|---|---|---|---|---|
| Trane | TTS730A | TW*739E15-C | 33,200 | 16.90 |
| Trane | TTX730A | TW*739E15+BAY24X045 | 32,000 | 13.45 |
| Trane | TTX730A | TW*739E15-C | 32,000 | 13.45 |
| Coleman | 9430G911 | 9435E833K | 29,600 | 13.20 |
| Coleman | 9430G911 | 3736-823K | 29,600 | 13.20 |

### *Cooling Capacity: approx 3 tons*

| Brand | Outdoor Unit | Indoor Unit | Capacity BTU/HR | Efficiency (SEER) |
|---|---|---|---|---|
| Trane | TTS73 6A | TW*739E15 | 39,000 | 16.20 |
| Lennox | HS14-411V-3P, 5P, 6P | CB15-41-1P | 41,500 | 15.00 |
| GMC | CTS36-1 | A50-XX+XV36+LL | 40,000 | 14.00 |
| Janitrol | CTS36-1 | A50-XX+EEP | 40,000 | 14.00 |
| Lennox | HS14-411V-3P, 5P, 6P | CH14-41-1FF | 38,000 | 13.30 |

## *Cooling Capacity: approx. 4 tons*

| Brand | Outdoor Unit | Indoor Unit | Capacity BTU/HR | Efficiency (SEER) |
|---|---|---|---|---|
| Lennox | HS14-511V-3P, 5P, 6P | CB15-65-1P | 52,000 | 14.00 |
| Lennox | HS14-511V-3P, 5P, 6P | CB15-46-1P | 49,500 | 13.20 |
| GMC | CTS48-1 | A65-XX+EEP | 50,000 | 13.00 |
| Janitrol | CTS48-1 | A65-XX+EEP | 50,000 | 13.00 |
| Lennox | HS14-511V-3P, 5P, 6P | CB15-41-1P+ LB53081CB | 49,500 | 13.00 |

Asterisk indicates a number or letter relating to model color or other nonessential feature.

Source: American Council for an Energy-Efficient Economy, 1989

## Central Heat Pumps

Heat pumps have two modes: heating and cooling. In the cooling mode, the heat pump works just like an air conditioner. In the heating mode, it works on the principle that even cool air contains some heat, and this heat may be extracted and concentrated, providing warm air to heat your home. In the heating mode, a heat pump uses a refrigerant to absorb heat from the cold outside air. A compressor concentrates the refrigerant, which causes the air's temperature to rise. The warmed air is then transferred indoors. Because they do not perform well over extended periods of subfreezing temperatures, heat pumps are generally most cost effective in regions without severe winters.

Like central air conditioners, the cooling ability of heat pumps is rated by a SEER; the heating ability is rated by the Heating Season Performance Factor, or HSPF, a ratio of the estimated seasonal heating output divided by the seasonal power consumption for an average U.S. climate. A typical new heat pump has an HSPF of about 6.5 and a SEER of about 9. The 1992 national energy-efficiency standards call for a minimum HSPF of 6.8 and a minimum SEER of 10.

# Central Heat Pumps

## Capacity: approx. 1.5 tons

| Brand | Outdoor Unit | Indoor Unit | Cooling Capacity BTU/HR | Cooling Efficiency SEER | Heating Capacity BTU/HR | Heating Efficiency HSPF |
|---|---|---|---|---|---|---|
| Coleman | 3718-611 | 3718-8** | 19,400 | 11.30 | 20,200 | 8.00 |
| Lennox | HP19-211-1P | CB19-21-1P+ LB-34792BE | 19,600 | 12.00 | 19,200 | 7.80 |
| Lennox | HP19-211-1P | CB1719-21-1P+ LB-34792BE | 19,600 | 12.00 | 19,200 | 7.80 |
| Lennox | HP19-211-1P | CB, CBH 19-26-1P+LB-34792 BE | 19,600 | 12.00 | 19,200 | 7.80 |
| Comfortmaker | CYD018G | DCYPAO18/024 | 18,000 | 10.30 | 17,200 | 7.30 |

## Capacity: approx. 2.0 tons

| Brand | Outdoor Unit | Indoor Unit | Cooling Capacity BTU/HR | Cooling Efficiency SEER | Heating Capacity BTU/HR | Heating Efficiency HSPF |
|---|---|---|---|---|---|---|
| Carrier | 38QE924300 | 38QE024300 | 24,000 | 14.70 | 24,000 | 9.20 |
| Coleman | 3724-911 | 3724-833 | 23,800 | 11.30 | 25,300 | 8.00 |
| Lennox | HP19-261-1P | CB19-31-1P+ LB34792BE | 26,000 | 11.55 | 25,200 | 7.90 |
| Lennox | HP19-261-1P | CBH19-31-1P+ LB-34792BE | 26,000 | 11.55 | 25,200 | 7.90 |
| American Standard | TWN724A | TWH739P15-C | 27,800 | 11.35 | 25,000 | 7.65 |

## Capacity: approx. 2.5 tons

| Brand | Outdoor Unit | Indoor Unit | Cooling Capacity BTU/HR | Cooling Efficiency SEER | Heating Capacity BTU/HR | Heating Efficiency HSPF |
|---|---|---|---|---|---|---|
| Coleman | 3730-901 | 3726-830 | 28,200 | 12.00 | 29,000 | 8.85 |
| Coleman | 3730-911 | 3726-8*3 | 28,200 | 12.00 | 29,000 | 8.85 |
| Trane | TWS730A | TW*739E15 | 31,800 | 16.40 | 29,600 | 8.70 |
| Rheem | RPGB-03JA | RHQAA-13+ RCQB-B030 | 30,200 | 12.20 | 29,400 | 8.40 |
| Ruud | UPGB-030JA | UHQA-13+ RCQB-B030 | 30,200 | 12.20 | 29,400 | 8.40 |

## *Capacity: approx. 3.0 tons*

| Brand | Outdoor Unit | Indoor Unit | Cooling Capacity BTU/HR | Cooling Efficiency SEER | Heating Capacity BTU/HR | Heating Efficiency HSPF |
|---|---|---|---|---|---|---|
| Carrier | 38QE936300 | 38QE036300 | 36,000 | 14.70 | 36,000 | 9.20 |
| Trane | TWS736A | TW*739E15 | 37,600 | 15.20 | 34,600 | 8.75 |
| Coleman | 3736-801 | 3736-830 | 37,000 | 11.30 | 38,500 | 8.50 |
| Coleman | 3736-811 | 3736-8*3 | 37,000 | 11.30 | 38,500 | 8.50 |

## *Capacity: approx. 3.5 tons*

| Brand | Outdoor Unit | Indoor Unit | Cooling Capacity BTU/HR | Cooling Efficiency SEER | Heating Capacity BTU/HR | Heating Efficiency HSPF |
|---|---|---|---|---|---|---|
| Coleman | 3742-911 | 3736-8*3 | 41,500 | 10.60 | 41,500 | 8.30 |
| Lennox | HP19-461-1P | CB19-51-1P+ LB34792BG | 43,000 | 11.50 | 43,000 | 8.30 |
| Lennox | HP19-461-1P | CH19-51+ B19 51-1P+ LB34792BG | 43,000 | 11.50 | 43,000 | 8.30 |
| Lennox | HP19-461-1P | CB19-51-1P | 43,000 | 10.50 | 43,000 | 8.30 |
| Lennox | HP19-461-1P | CH19-51-1+ B19 51-1P | 43,000 | 10.50 | 43,000 | 8.30 |

## *Capacity: approx. 4.0 tons*

| Brand | Outdoor Unit | Indoor Unit | Cooling Capacity BTU/HR | Cooling Efficiency SEER | Heating Capacity BTU/HR | Heating Efficiency HSPF |
|---|---|---|---|---|---|---|
| Rheem | RPGB-048JA | RHQA-16+ RCQB-B048 | 47,500 | 11.25 | 49,000 | 8.70 |
| Ruud | UPGB-048JA | UHQA-16+ RCQB-B048 | 47,500 | 11.25 | 49,000 | 8.70 |
| Rheem | RPGB-048JA | RCQB-B048 | 47,000 | 10.95 | 49,500 | 8.60 |
| Ruud | UPGB-048JA | RCQB-B049 | 47,000 | 10.95 | 49,500 | 8.60 |
| Lennox | HP14-311/511V-10, 12P | CB, CBH 19-51-1P+LB34792BF | 48,500 | 10.80 | 47,500 | 8.15 |

Asterisk indicates a number or letter relating to model color or other nonessential feature.

Source: American Council for an Energy-Efficient Economy, 1989

## Gas Furnaces

New furnaces are rated by their Annual Fuel Utilization Efficiency, or AFUE, a measure of overall seasonal performance. The average new gas furnace sold in 1988 had an AFUE of 75 percent. The 1992 national efficiency standards call for a minimum AFUE of 78 percent.

The efficiency of a gas furnace has a lot to do with the sophistication of the equipment. There are five major systems:

| Type of furnace | Typical AFUE |
|---|---|
| Gas pilot light | 62% |
| Electronic ignition | 68% |
| Automatic vent damper | 76% |
| Power combustion | 82% |
| Condensing furnace | 90+% |

Condensing furnaces are often priced much higher than less-efficient types of furnaces, making them most economical in areas with long, cold winters. Those in regions with moderate or mild winters may suffice with an AFUE in the 78 percent to 85 percent range. Of course, one important energy-efficiency features is to make sure that the furnace capacity is appropriate for your home.

### *Gas Furnaces*

### *25,000–45,000 BTU/HR Heating Capacity*

| Brand | Model | BTU/HR | Efficiency (% AFUE) |
|---|---|---|---|
| Bryant, Payne | 398AAW030040 | 40,000 | 97.3 |
| Carrier | 58SX040-BC | 40,000 | 97.3 |
| Carrier | 58SX040-FG | 40,000 | 97.3 |
| Day and Night | 398AW030040 | 40,000 | 97.3 |
| Armstrong | EG6E40DC13 | 39,000 | 97.1 |

## *40,000–60,000 BTU/HR Heating Capacity*

| Brand | Model | BTU/HR | Efficiency (% AFUE) |
|---|---|---|---|
| Bryant, Payne | 398AAW030040 | 40,000 | 97.3 |
| Carrier | 58SX040-BC | 40,000 | 97.3 |
| Carrier | 58SX040-FG | 40,000 | 97.3 |
| Day and Night | 398AW030040 | 40,000 | 97.3 |
| Armstrong | EG6E40DC13 | 39,000 | 97.1 |

## *61,100–80,000 BTU/HR Heating Capacity*

| Brand | Model | BTU/HR | Efficiency (% AFUE) |
|---|---|---|---|
| Clare Brothers | HEGB 80S | 77,000 | 96.0 |
| Clare Brothers | HEHB 80-S | 77,000 | 95.1 |
| Amana | EGHW100DC3 | 76,000 | 95.0 |
| Armstrong | EG6E80DC16, 19 | 76,000 | 94.7 |
| Glowcore | UGR080D1* | 76,000 | 94.7 |

## *81,000–105,000 BTU/HR Heating Capacity*

| Brand | Model | BTU/HR | Efficiency (% AFUE) |
|---|---|---|---|
| Clare Brothers | HEGB100S | 96,000 | 96.0 |
| Airco | AHC52-90 | 85,100 | 95.0 |
| Duomatic Olsen | HCS2-90 | 85,100 | 95.0 |
| Lennox | G14Q4/5-100- | 95,000 | 95.0 |
| Amana | EGHW100DC3 | 95,000 | 94.5 |

## 106,000–135,000 BTU/HR Heating Capacity

| Brand | Model | BTU/HR | Efficiency (% AFUE) |
|---|---|---|---|
| Clare Brothers | HEG-B120 | 112,000 | 93.4 |
| Airco | HCS 120(M) | 113,600 | 93.3 |
| Duomatic Olsen | HCS 120(M) | 113,600 | 93.3 |
| Central Env. Sys. | P1UDD20N13301A | 133,000 | 93.0 |
| Clare Brothers | HEMB-B130 | 123,000 | 92.9 |

## Gas Boilers (Hot Water)
## 40,000–60,000 BTU/HR Heating Capacity

| Brand | Model | BTU/HR | Efficiency (% AFUE) |
|---|---|---|---|
| Hydrotherm | A-50B | 47,000 | 90.4 |
| Glowcore | GB060-7 | 53,460 | 89.0 |
| Buderus Logana | 105/15 | 52,000 | 87.4 |
| H. B. Smith | G901-W-5HS1 | 57,300 | 87.3 |
| Heatmaker/Trianco | 60H** | 52,000 | 87.0 |

## 61,000–90,000 BTU/HR Heating Capacity

| Brand | Model | BTU/HR | Efficiency (% AFUE) |
|---|---|---|---|
| Hydrotherm | A-100B | 88,000 | 90.4 |
| Glowcore | GB090-1 | 79,900 | 88.8 |
| Buderus Logana | 105/21 | 72,000 | 87.7 |
| Weil-McLean | VHE-4 | 87,000 | 87.1 |
| Ultimate | G-7-D | 73,000 | 85.5 |

## *91,000–150,000 BTU/HR Heating Capacity*

| Brand | Model | BTU/HR | Efficiency (% AFUE) |
|---|---|---|---|
| Hydrotherm | A-150 B,C | 132,000 | 90.6 |
| Glowcore | GB16014 | 142,0000 | 88.8 |
| Buderus Logana | 105/28 | 96,000 | 87.8 |
| H. B. Smith | G901-W-7HS1 | 95,700 | 87.7 |
| Buderus Logana | 205/42 | 143,000 | 87.4 |

## *Gas Boilers (Steam)*

| Brand | Model | BTU/HR | Efficiency (% AFUE) |
|---|---|---|---|
| Burnham | XG-4004 | 88,000 | 83.5 |
| Burnham | XG-4005 | 117,000 | 83.4 |
| Burnham | XG-4006 | 147,000 | 83.3 |
| Weil-McLean | EG-30-PID | 63,000 | 82.7 |
| Columbia | CFG-75AID | 61,000 | 81.4 |

Asterisk indicates a number or letter relating to model color or other nonessential feature.

Source: American Council for an Energy-Efficient Economy, 1989

## Oil Furnaces

The 1992 national efficiency standards for oil furnaces are the same as for gas furnaces: a minimum of 78 percent. High-efficiency oil furnaces achieve their high ratings through automatic flue dampers and "flame retention" burners.

## *Oil Furnaces*

### *60,000-80,000 BTU/HR Heating Capacity*

| Brand | Model | BTU/HR | Efficiency (% AFUE) |
|---|---|---|---|
| Yukon | 85-100-1500 | 59,000 | 90.6 |
| Yukon | H-70-0-02 | 64,000 | 89.5 |
| Yukon | H-70-0-02 | 63,000 | 87.8 |
| Airco | AHTL 80B | 78,000 | 87.7 |
| Duomatic Olsen | HTL 80B | 78,000 | 87.7 |

## *85,000–104,000 BTU/HR Heating Capacity*

| Brand | Model | BTU/HR | Efficiency (% AFUE) |
|---|---|---|---|
| Yukon | U-90-0-03 | 85,000 | 90.8 |
| Airco | ABCL 90 | 90,000 | 87.8 |
| Duomatic Olsen | BCL 90 | 90,000 | 87.8 |
| Airco | AHTL 90B | 89,000 | 87.3 |
| Duomatic Olsen | HTL 90B | 89,000 | 87.3 |

## *105,000–135,000 BTU/HR Heating Capacity*

| Brand | Model | BTU/HR | Efficiency (% AFUE) |
|---|---|---|---|
| Energy Kinetics | System 2000 | 120,000 | 87.0 |
| Williamson | 1454-14 | 118,000 | 87.0 |
| Williamson | R164-15-3 | 119,000 | 86.8 |
| Airco | ABCL 115 | 118,000 | 86.7 |
| Duomatic Olsen | BCL 115 | 118,000 | 86.7 |

# *Oil Boilers (Hot Water)*

## *60,000–84,000 BTU/HR Heating Capacity*

| Brand | Model | BTU/HR | Efficiency (% AFUE) |
|---|---|---|---|
| Axeman-Anderson | OL-91 | 80,000 | 88.1 |
| Ultimate | K-5T-D | 61,700 | 88.0 |
| Buderus Logana | 105/21 | 72,000 | 87.7 |
| Ultimate | K-5-D | 73,900 | 87.6 |
| Ultimate | PFO-4 | 74,000 | 87.6 |

## *85,000–104,000 BTU/HR Heating Capacity*

| Brand | Model | BTU/HR | Efficiency (% AFUE) |
|---|---|---|---|
| Dynatherm | MXA | 104,000 | 89.1 |
| Dynatherm | MX*-A | 104,000 | 89.1 |
| Buderus Logana | 105/28 | 96,000 | 87.8 |
| Crown | CTPR-3 | 92,000 | 87.6 |
| Ultimate | PFO-5 | 93,000 | 87.6 |

## *105,000–134,000 BTU/HR Heating Capacity*

| Brand | Model | BTU/HR | Efficiency (% AFUE) |
|---|---|---|---|
| Axeman-Anderson | 74NPO-V | 105,000 | 88.7 |
| Ultimate | PFO-7 | 130,000 | 87.6 |
| Crown | CTPR-4 | 130,000 | 87.5 |
| Ultimate | PFO-6T | 117,000 | 87.5 |
| Buderus Logana | 205/34 | 116,000 | 87.2 |

## *135,000–175,000 BTU/HR Heating Capacity*

| Brand | Model | BTU/HR | Efficiency (% AFUE) |
|---|---|---|---|
| Buderus Logana | 205/50 | 171,000 | 87.6 |
| Ultimate | PFO-8T | 154,000 | 87.5 |
| Axeman-Anderson | 108 NPO-U | 153,000 | 87.4 |
| Buderus Logana | 205/42 | 143,000 | 87.4 |
| Crown | CTPR-5 | 167,000 | 87.4 |

## Oil Boilers (Steam)

### 61,000-91,000 BTU/HR Heating Capacity

| Brand | Model | BTU/HR | Efficiency (% AFUE) |
|---|---|---|---|
| Utica | SF-365 | 79,000 | 86.0 |
| H. B. Smith | 8-S/W-3L | 91,000 | 85.8 |
| Peerless | JO/JOT-TW075-S | 91,000 | 84.2 |
| Utica | SF-365SD | 78,000 | 84.0 |

### 92,000-150,000 BTU/HR Heating Capacity

| Brand | Model | BTU/HR | Efficiency (% AFUE) |
|---|---|---|---|
| H. B. Smith | 8-S/W-4L | 131,000 | 85.9 |
| H. B. Smith | 8-S/W-3H | 102,000 | 85.6 |
| Slant Fin | L30PZ | 134,000 | 84.1 |
| Burnham | V-74S | 125,000 | 82.9 |

Asterisk indicates a number or letter relating to model color or other nonessential feature.

Source: American Council for an Energy-Efficient Economy, 1989

## Products to Save Energy

In addition to major appliances and lighting, here are several smaller, inexpensive products that can save you energy and money.

**Air-conditioner covers:** If you cannot remove your room air conditioner from the window during the winter, consider covering it, both inside and out. Besides protecting your air-conditioning unit, these covers also help keep cold air from entering your home through the space around the air conditioner thus cutting down on heating costs. One brand is **Frost-King Air-Conditioner Cover**.

**Caulking:** Filling in the small spaces and gaps around window openings and where pipes and wires enter the home reduces drafts that reduce the efficiency of your heating and air-conditioning system. Most caulking products cost under $10; rope caulk, one of the easiest types to apply, sells for about $4 for 40 or 50 feet. There are many brands available, sold at most hardware stores, including **DAP Rely-ON, Dow-Corning Silicone Caulk, Easy Caulker Acrylic Foam,** and **Mortite Rope Caulk**.

**Draft Blockers:** These foam plates fit behind light switches and electrical outlets to reduce drafts that come in through those spaces. You can get a packet of 10 for about $3; they are easy to install, with only a screwdriver.

**Draft Guards:** These are sand- or pebble-filled tubes that you place around windowsills and sashes to reduce the loss of heated and cooled air. (If you prefer to sleep with the window open in winter, placing a draft guard under your bedroom door will prevent the cold air from seeping into the rest of the house.) Draft guards cost about $7 and are sold under a variety of names, including **Frost-King Stop Draft Weatherstrip**.

**Heat Reflectors:** These are thin sheets (usually foam coated on one side with metal foil) that fit behind radiators, to reflect heat away from the wall and into the room, thereby maximizing each radiator's efficiency. The **Viking Heat Reflector** is one such product.

### Do Microwave Ovens Save Energy?

Are microwave ovens energy-efficient? Yes, if you use them in the right way. Microwaves reduce electricity costs by 25 to 50 percent when used instead of an oven for cooking purposes. However, when used for defrosting microwaves will increase electricity use. It is far more energy efficient to defrost frozen foods in the refrigerator or on a countertop.

**Programmable Thermostat:** These new types of thermostats allow you to change the temperature (of both heating and air conditioning) at different times of day. You might, for example, have the heater go to 68 degrees at 6 A.M. (to have everything warm by the time you get up), drop to 60 degrees at 9:30 A.M. (after everyone has left for the day), go back up to 68 degrees at 4 P.M. so the house will be warm when you return for the evening), then drop back to 60 degrees at 11 P.M. (while you are in bed). Some programmable thermostats also have a second set of settings for weekends, when people are often at home more during the day. The thermostats, which range from $90 to $175, are not cheap, but they can save up to 12 percent on your heating bill each year. The device surely will pay for itself within two or three years. There are many brands, available at most hardware stores, including **White-Rogers Comfort-Set, Sears Thrift-T, Honeywell Fuel Saver,** and **Weather Wizard.**

**Reflective Window Film:** These are thin plastic sheets you place directly on the inside of window panes and glass doors. The film reflects inside heat back into your home, reducing the amount that is conducted outside through windows. The film, which costs about $10 a window, is easy to put on; it adheres to the window directly, or with the help of water from a spray bottle. One good brand is **3M Scotchtint Reusable Sun Control Film.**

**Storm Window Kits:** It can be expensive to have storm windows installed throughout your house (although the investment will probably pay for itself in fuel savings and increased resale value of your home), but there is a less-expensive alternative. Storm window kits consist of plastic film or sheets to cover the window. Attaching the plastic to the window is done with tape or tacks. One product, **3M Window Insulation Kit**, includes "shrink" film that you apply by using the hot air from a hair dryer to tighten the film and shrink away wrinkles. Other brands include **Frost-King Storm Window Kit, Sears Window Insulating Kit,** and **TYZ All-Weather Window**. Prices range from about $3 to $10 per window.

**Weatherstripping:** This includes plastic, foam, felt, or rubber strips that fit around window and door frames to create a tight seal and reduce heated and cooled air from escaping outside. Most are easy to apply, usually by way of a self-adhesive backing. Prices vary, but average about $5 per window or door. Brands include **Frost-King Weatherstrip Tape, 3M V Seal Weatherstrip, Jalousie Window Weatherstrip, Stix-on Self-Adhesive Weatherstripping,** and **Mortite Weatherstripping and Caulking Cord**.

## Saving Water at Home

We waste a lot of water, especially hot water. This wasn't always a problem, because there was plenty of water to go around. That's no longer the case, because of increasing populations, the greater likelihood of droughts resulting from global warming, more water pollution, and acid rain. All of this means that surpluses of water will increasingly be a thing of the past, and water shortages will become routine.

The typical American home uses almost 300 gallons of water a day. The fact is, you can save a great deal of that without making major sacrifices, and save money in the process in the form of lower sewer, water, and energy bills. If you have a septic tank—30 percent of the population does—conserving water will reduce the wear and tear on your system, and will require less energy from pumping well water. And then there are the tax dollars saved by not having to expand existing water treatment plants or to build new ones.

How do we waste water? Most of it is wasted around the home. A steady drip on a faucet can waste 20 gallons of water a day. A leaking toilet can waste 200 gallons a day without making a sound. At the water pressure found in most household plumbing systems, a 1/32" leak in a faucet can waste up to 6,000 gallons a month, or 72,000 gallons a year.

Fortunately, there are a growing number of appliances and bathroom fixtures on the market that use considerably less water. In fact, it is likely that the water-saving features of many of these devices will be required in coming years. In 1989, Massachusetts became the first state to require that all toilets sold use no more than 1.6 gallons per flush. (Most toilets manufactured before 1982 use six to seven gallons per flush; since 1982, most toilets sold in the United States operate at a flow of 3.5 gallons per flush under voluntary industry standards.) The federal government is considering a nationwide law similar to that of Massachusetts, which would also cover shower heads and faucets.

## Ten Steps for Saving Water

Here are ten simple steps you can take to minimize the amount of water you waste at home:

**1.** Test for a leaking toilet by adding food coloring to the tank (not the bowl; the tank is the covered portion of the toilet that contains the flushing mechanism). Without flushing, note if any color appears in the bowl after 30 minutes. If color appears, you have a leak.
**2.** Check your water meter while no water is being used. If the dial moves, you have a leak.
**3.** Turn off your water and hot water heater when going on a long trip.
**4.** When washing dishes, don't run water continuously in the sink. If you use a dishwasher, run it only when you have a full load and use the cycles with the fewest number of washes and rinses.
**5.** Add your garbage to the compost or trash instead of putting it down the disposal. Disposals not only use a great deal of water, they also add solids to an already overloaded sewer system.
**6.** Use a toilet dam to save water every time you flush. *Do not—repeat, do not—put a brick in your toilet tank to save water. The brick will eventually wear away and sediment may clog your pipes.* Most commercial dams are very inexpensive and can save up to 4 gallons per

flush. Alternatively, fill gallon, half-gallon, or quart jugs with water and set them inside the tank. (This is an appropriate use for those plastic jugs you've already purchased but don't want to throw away.)

**7.** Install a low-flow shower head. It will reduce water use by up to a third without affecting the shower pattern. A normal shower uses about eight gallons of water per minute. Low-flow devices can cut the flow to 2 or 2 1/2 gallons per minute.

**8.** Install aerators on your faucets. These screw onto the faucet and add air to the water, giving a fuller flow with less water—about 40 to 60 percent less water. They also reduce splashing. Aerators are available at any hardware store for a dollar or two.

**9.** Water your lawn only during the coolest part of the day to avoid rapid evaporation.

**10.** Take showers instead of baths. A bath can use 30 to 50 gallons of water, compared to 20 gallons for a 10-minute shower with a low-flow shower head.

**Where to Get Water-Saving Devices.** Many of the products described below can be found in local hardware stores or the houseware departments of department stores. Or contact the manufacturers and distributors below for their latest catalogs.

❑ **Co-op America** (2100 M St. NW, Washington, DC 20036; 202-872-5307, 800-456-1177) offers several water-saving devices in its catalog, including Europa Showerheads ($11.75 to $19) in a range of colors, from "china blue" to "cherry red" to "brite white."

❑ **Ecological Water Products** (1341 W. Main Rd., Middletown, RI 02840; 401-849-4004) sells lower-flow shower heads and other water-saving devices.

❑ **Resources Conservation, Inc.** (P.O. Box 71, Greenwich, CT 06838; 203-964-0600, 800-243-2862) distributes a line of "Drought Buster" water-saving devices sold in hardware stores and home centers, including shower heads, faucet aerators, and toilet dams.

❑ **Seventh Generation** (10 Farrell St., South Burlington, VT 05403; 802-658-5507, 800-456-1177) sells several water-saving devices, including a low-flow shower head ($13.95) and toilet dams ($6.50, three for $18.50). The "Water Saver Kit" contains two shower heads, two toilet dams, and three faucet aerators for $59.95.

## ENERGY-EFFICIENT LIGHTING

More than 20 percent of all the electricity generated in the United States is used for lighting. Half of that energy is wasted—used to light unoccupied rooms or squandered on inefficient lighting sources. It used to be that cutting lighting energy use consisted of replacing your light bulbs with lower-wattage ones, or remembering to turn off lights when you left the room. No more. There is a host of innovative products that actually can provide *more* light for less energy, less pollution—and less money.

How much pollution is caused by inefficient lighting? Estimates vary, but several estimates put the reduced carbon dioxide emissions from a typical coal-fired power plant resulting from replacing a single standard incandescent bulb with a compact fluorescent bulb at around one ton over the bulb's lifetime. Carbon dioxide is a major contributor to global warning, and power plants in general contribute to other forms of air and water pollution, including acid rain. According to the Lawrence Berkeley Laboratory, here are the estimated savings resulting from replacing 800 million incandescent bulbs with compact fluorescent ones:

| | |
|---|---|
| Net annual savings: | $6 billion |
| Unburned coal: | 30 metric tons |
| Unneeded power plants: | 12 1,000 megawatt plants |
| Unneeded oil platforms: | 30 platforms producing 10,000 barrels a day |

As with other household "appliances," you probably don't realize that the cost of a light bulb is only a fraction of its total cost. The "operating cost"—the amount you pay for electricity—is five to ten times the cost of the bulb itself. So, those two bulbs you are buying for $2.50 actually cost you $12.50 to $25 in electricity.

To understand how to save money, it may first help to understand the three basic types of household lighting.

❑ **Incandescent:** This type—the standard light bulb invented in the 1800s—is the most common lighting source in American homes. It is also the most wasteful. Ninety percent of the energy consumed by an incandescent lamp is given off as heat rather than visible light. This is because incandescents do not directly convert electricity to light. Rather, they use electricity to heat a coiled filament in a vacuum or inert gas-filled bulb, making the filament glow.

❑ **Fluorescent:** Fluorescent lamps—usually sold as long, thin tubes with two prongs at each end—convert electricity to visible light by using an electric charge to "excite" gaseous atoms. These atoms emit ultraviolet radiation, which is absorbed by the phosphor coating on the tube walls. The phosphor coating "fluoresces," producing a visible light. Fluorescent lamps convert electricity to visible light up to five times more efficiently than incandescent lamps, and last up to twenty times longer. Most fluorescent lamps require a special fixture, although a new generation of compact fluorescent lamps can fit in existing "screw-in" fixtures.

❑ **High-intensity discharge (HID):** The most common type found in homes is halogen, found increasingly in reading lights and some overhead fixtures. These lamps vary from the standard incandescent in that a halogen gas is added to the bulb. Halogen gas reduces the filament evaporation rate, thereby reducing the rate at which the glass bulb darkens. This increases lamp life up to four times that of a standard bulb. Moreover, tungsten halogen lamps emit more light for the same amount of electricity. These bulbs require special fixtures, however, and can become very, very hot while lit. The bulbs are also very expensive compared to standard incandescent or fluorescent bulbs.

**Compact fluorescents.** The most promising technology for Green Consumers is the new compact fluorescent bulb. An 18-watt screw-in fluorescent bulb produces as much light as a 60- or 75-watt incandescent bulb and lasts ten times longer. Unlike fluorescent tubes, compact bulbs don't hum or flicker, and because of improved color rendering they produce light comparable to the incandescent variety. Unfortunately, many people are put off by the $12-to-$18-per-bulb price tag. But that is shortsighted: because the bulbs last much longer and use about 90 percent less electricity, compact fluorescent bulbs are far cheaper over their life cycle. A single bulb will *save* you $25 to $40 in electricity over the life of the bulb. That makes for an impressive investment, and an environmentally sound one.

Note, however, that compact fluorescent bulbs are not recommended for enclosed fixtures, fixtures connected to dimmer switches, or outdoor use.

Compact fluorescent bulbs are manufactured by **Philips Lighting Co.** (200 Franklin Sq., Somerset, NJ 08875; 201-563-3000, 800-543-8167); **General Electric Co.** (Lighting Information Center, Nela Park, Cleveland, OH 44112; 800-626-2000), **Satco Products, Inc.** (110 Heartland Blvd., Brentwood, NY 11717; 516-243-2022), and **Osram** (110 Bracken Rd., Montgomery, NY 12549; 914-457-4040; 800-431-9980). If you cannot find them in your local hardware or lighting supply store, here are three other sources:

❑ **Seventh Generation** (10 Farrell St., South Burlington, VT 05403; 802-658-5507, 800-456-1177) sells a variety of compact fluorescent bulbs and fixtures.
❑ **Jade Mountain** (P.O. Box 4616, Boulder, CO 80306; 303-449-6601) sells various sizes of Philips's bulb, along with other energy-saving lighting fixtures.
❑ **White Electric Co., Inc.** (1511 San Pablo Ave., Berkeley, CA 94702; 415-845-8534, 800-468-2852) sells bulbs from Satco and Philips. Minimum order is 12 bulbs, but prices are lower than many other retailers.

## Home Fire Extinguishers

Compact fire extinguishers, the kind most often sold for home use, usually contain halon-211, which have 10 times the ozone depletion potential of the most damaging CFCs. For halon-free extinguishers, ask for the dry-chemical type, which usually contain sodium carbonate. Halon-free extinguishers intended for household use are available from **Ansul Fire Protection** (1 Stanton St., Marinette, WI 54143; 715-735-7411, 800-346-3626), under the "Sentry" and "Red-Line" brands, and from **Sears**, under the "Craftsman" brand. Note, however, that not all models marketed under each of these brand names are halon-free.

## Good Wood

The destruction of the tropical rain forests is one of the most serious environmental catastrophes of our time. (See "Seven Environmental Problems You Can Do Something About.") Each year, according to the World Bank and the United Nations Development Programme, at least 12.5 million acres of rain forest are destroyed by commercial loggers. **Mahogany, teak, ramin, lauan,** and **meranti** are some of the timbers extracted for trade. These hardwoods are used for many purposes, from furniture to a great many do-it-yourself projects. The problem is that once these rain forests have been logged, they are often so badly damaged that the remaining trees will not grow properly. While only a few tree species have commercial value, as much as two-thirds of the forest is destroyed in the logging process. Estimates are that for every tree felled, another ten may be destroyed or damaged beyond hope. The average Southeast Asian logging concession wastes over one and a quarter cubic yards of wood for every cubic yard of useful timber removed from the forest. Several tree species, such as the mahoganies from Cuba and Africa, have become commercially extinct, while others are increasingly scarce. Several tropical countries that once maintained lush forests have depleted them to the point that they now must import wood, according to **Rainforest Action Network**, one of the leading groups on this issue. An increased demand for inexpensive third world lumber (due mostly from lower wages, government subsidies, and poor forestry policies that encourage short-term financial returns) have resulted in the destruction of many ecologically important forests.

How is this wood used? There are two principal ways:

❑ Most U.S. imports of tropical hardwood are in the form of **plywood** and **paneling** from Southeast Asia, but sawn timber (lumber) is also imported, especially Brazilian mahogany. Much plywood is also finished in the U.S.—that is, the thin sheets that make up the plywood are imported, then glued together. Most plywood uses timber from the family of trees known as the dipterocarps, especially the shorea species, known commercially as **lauan, meranti**, or **Philippine mahogany**.

## *Treated Wood*

Pressure-treated wood is sold for a wide variety of home, farm, commercial, and industrial uses. Pressure-treated wood resists insect attack and decay because it has been treated with chemicals—usually arsenic compounds, creosote, or pentachlorophenol (penta). However effective, these chemicals are very toxic, and if not handled properly can be hazardous to your health. Aside from harming human health, the chemicals also harm the environment. Nearly 40 penta waste-treatment plants are on the Environmental Protection Agency's national Superfund list for priority cleanup, according to Environmental Action.

Here are some suggestions for safe use of treated woods:

❑ First and foremost, read all labels.
o Don't breathe sawdust from wood treated with these preservatives. Wear a dust mask when sawing.
❑ Protect your eyes by wearing goggles.
❑ Avoid skin contact when working with penta- or creosote-treated wood. Wear tightly woven coveralls and gloves made of impervious materials.
❑ Wash exposed skin thoroughly after handling wood treated with any of these preservatives, especially before eating, drinking, or smoking.
❑ Wash garments that have residues or sawdust from these chemicals separately from other laundry.

Treated wood has its uses in construction, but it should be used sparingly and with care.

## *North American Alternative Woods*

| Wood | Furniture | Cabi-netry | Millwork | Veneer | Gen. Const. | Boxes crates | Flooring | Plywood |
|---|---|---|---|---|---|---|---|---|
| Ash | ● | ● | | | | | | |
| Basswood | | | ● | I [a] | | | | |
| Beech | ● | | ● | ● [a] | | ● | ● | |
| Birch | ● | ● | ● | ● [a] | | | ● | ● |
| Butternut | ● | ● | ● | | | ● | | |
| Cherry | ● | ● | ● | ● [c] | | | | |
| Cottonwood | | | | ● [a] | | ● | | ● |
| Cypress | ● | ● | ● | | | | | |
| Elm | ● | | ● | | | | | |
| Douglas fir | | | ● | | ● | | | |
| Black gum | ● | | ● | ● [b] | | ● | ● | |
| Red gum | ● | ● | ● | ● [c] | | | | ● |
| Hackberry | ● | | ● | | | | | |
| Western hemlock | | | | | ● | ● | ● | |
| Hickory | | | | | | | | |
| Sugar maple | ● | | ● | | | | ● | |
| Soft maple | ● | | | | | ● | | |
| Red oak | ● | ● | ● | | | | ● | |
| White oak | ● | ● | ● | | ● | | ● | |
| Pecan | ● | | | | | | ● | |
| Ponderosa pine | ● | ● | ● | | ● | | | ● |
| Yellow Southern pine | | | ● | | ● | | | |
| Yellow poplar | ● | ● | ● | | | | | ● |
| Sitka spruce | | | ● | | ● | | | |
| Sycamore | ● | ● | ● | ● [a,b] | | ● | | |
| Black walnut | ● | ● | ● | ● [c] | | | | |

**Explanation of some categories:**

*Veneer* (a) used in the manufacture of plywood, for concealed parts of furniture; (b) used in crating and container plywood veneers; (c) face veneer for furniture and interior grade plywood.

*Millwork*: interior trim, sash, door, molding, and paneling.

*General Construction*: framing, sheathing, roofing, and subflooring.

Source: Rainforest Action Network

❑ Other major tropical hardwood imports are in the form of **ready-made furniture** or **furniture components** from manufacturing countries in Southeast Asia. Exotic tropical timbers such as ebony and rosewood may also be used for craft work and special items such as musical instruments. Teak from Southeast Asia is commonly used by boat builders.

What can Green Consumers do about the tropical hardwood problem? Unfortunately, when buying lumber or finished products, it is often difficult to tell the type of wood or its country of origin. The chart on page 213 will help you determine how many of these woods are used. Your best bet is to ask questions of store owners to find out what tropical hardwoods they import, and to let them know that you don't plan to buy products that contain them.

## FURNITURE

When buying furniture, the most important consideration is whether it is made of tropical hardwoods. But also consider what's contained *inside* the furniture. Many, many furniture cushions are made from plastic foam, almost all of which contains chlorofluorocarbons, which damage the atmospheric ozone layer (see "Seven Environmental Problems You Can Do Something About"), or methylene chloride, another blowing agent used by some foam manufacturers, but which has been listed as a "probable" carcinogen by the Environmental Protection Agency. However, three furniture foam products introduced in 1989 by **Union Carbide**— "Geolite" (used in cushioning for the arms and backs of upholstered furniture), "Hyperlite" (a molded foam cushioning for use in automobiles), and "Ultracel" (for use in furniture seat cushioning)—contain no CFCs. Unfortunately, Union Carbide would not release a list of products containing their foam cushioning products, but if you buy furniture containing foam, you should consider asking for it. Some products containing Ultracel are so indicated on product tags, but some manufacturers don't indicate this.

## Safer Household Paints

We rarely consider the environment when buying paint, but perhaps we should. Many commercial paint products are manu-

### *Paint Hazardous-Waste Disposal*

| Chemical Products | Hazardous Ingredients | Alternatives | Hazard Properties | Disposal |
|---|---|---|---|---|
| Enamel or oil based | pigments, ethylene, aliphatic hydrocarbons, mineral spirits | latex or water based paint | flammable, toxic | 2, 1 |
| Furniture strippers | acetone, methyl ethyl ketone, alcohols, xylene, toluene, methylene, chloride | sandpaper or heat gun | flammable, toxic | 2 |
| Latex or water based | resins, glycol ethers, esthers, pigments, phenyl mercuric acetate | limestone-based white wash or cassein-based paint | toxic | 1, 4 |
| Rust paints | methylene chloride, petroleum distillates, toluene | unknown | flammable, toxic | 2 |
| Stains and finishes | mineral spirits, glycol ethers, ketones, halogenated hydrocarbons, naphtha | latex paint or natural earth pigment finishes | flammable, toxic | 2 |
| Thinners and turpentine | n-butyl alcohol, acetone, methyl isobutyl ketone, petroleum distillates | use water with water based paints | flammable, toxic | 3 |
| Wood preservatives | chlorinated phenols (e.g., PCP), copper or zinc, naphthenate, creosote, magnesium fluorosilicate | water based wood preservatives | flammable, toxic | 2 |

**Disposal Notes**

1. Recyclable. Take to a service station, reclamation center, or household hazardous-waste collection. Partially used useful products such as paints may be exchanged.

2. These wastes should be safely stored until a hazardous-waste program is organized in your community. Fully spent currently available pesticide containers may, however, be triple rinsed and the rinse water poured down the drain and flushed with water.

3 . Keep in tightly closed jar and allow contaminants to settle out. Strain the thinner through a fine mesh sieve; reuse the liquid. The concentrated contaminants should be stored and taken to a collection center.

4. Fully use these products. Air-dry latex paints and discard container in trash.

Courtesy Environmental Hazards Management Institute, 10 Newmarket Rd., P.O Box 932, Durham, NH 03824; 603-868-1496

factured from nonrenewable materials, such as crude oil and coal-tar oil. To turn them into paints requires a variety of toxic chemicals and a lot of energy. To stabilize some of the chemicals may require the addition of still other chemicals. Each step requires the use of great quantities of water, some of which will end up full of these chemicals' by-products, and which must be disposed of in some way.

Fortunately, there is a growing number of paints made from natural materials, in particular plant products, which can be manufactured by much safer processes. Forgoing chemicals in favor of beeswax, carnauba wax, plant extracts, and plant gums, these manufacturers are creating high-quality, waterproof paints that rival their traditional counterparts.

Here are three sources of natural paint products:

❑ **AFM Enterprises** (1140 Stacy Ct., Riverside, CA 93507; 714-781-6860) manfactures water-based paints, lacquers, glues, and waxes.
❑ **Auro Organic Paints** (Sinan Co. Natural Building Materials, P.O. Box 181, Suisun City, CA 94585; 707-427-2325) manufactures a line of natural paints, varnishes, waxes, lacquers, glues, cleansers, and polishes, all made with what the company calls its "gentle chemistry." As with Livos, most products are made with beeswax, natural oils, chalk, plants, and other natural ingredients.
❑ **Livos PlantChemistry** (2641 Cerrillos Rd., Santa Fe, NM 87501; 505-988-9111) manufactures a wide range of paints, oil finishes, shellacs, stains, varnishes, spackels, adhesives, thinners, waxes, and polishes, all made from natural ingredients. All Livos products are made from earth pigments and are "cruelty free," meaning that no animal testing or animal ingredients are used, with the exception of beeswax. Livos claims that its products can be used safely on "food-related items such as cutting boards, wooden bowls and plates, kitchen counters, and butcher blocks," as well as on most other paintable surfaces.

# P·E·R·S·O·N·A·L C·A·R·E P·R·O·D·U·C·T·S

While the cosmetics and personal care products industries have long been concerned with the ecologies of our bodies, they have only recently begun considering that of our planet. The past two decades have seen a remarkable growth in products intended to make us feel good while they make us look good, too. (Indeed, a good case can be made that the two go hand in hand.) But whereas yesterday's products took on a look-good-whatever-it-takes attitude, many of today's products place a premium on what effects they might have on our short- and long-term health.

What about the environment? There's some progress there, too, but not much. One of the biggest problems is packaging—the creators of cosmetics and health products are masters of it. From the glorious shapes of bottles to the convenience of pocket- and purse-size objects, the purveyors of everything from shampoos to shaving cream to suntan oil have tried to accommodate our every need. And in the process, they have contributed significantly to the mountain of waste we create, and to some of the harmful matter this trash releases into our air, water, and soil.

Another issue for Green Consumers involves the process involved in getting many of these products onto our shelves. Throughout the world, there has been growing concern about the use of animals in the safety testing of all types of products. Their use in cosmetics has been particularly controversial because these products are considered nonessential. To address this increased concern, a growing number of companies have produced what have come to be called "cruelty-free" products—those that do not involve animal testing or that use animals in a responsible manner. (See page 224 for more information about cruelty-free products.)

The good news is that, on all fronts, Green Consumers have an increasing number of choices available of quality products that don't harm the earth—or any of its creatures.

## – The Dangerous Convenience of Disposables –

We've long been called the "throw-away society," and with good reason: nearly everything we buy, it seems, was meant to be used once or twice and thrown away. It wasn't that long ago that "built to last" was a favorite phrase of advertisers. (It still is for things like cars and washing machines, but little else.) Now, "use it and lose it" might be a more fitting slogan. Unfortunately, despite their intended short life span, many of these disposable products are still "built to last"—at least in terms of the time it takes for their packages to disappear from our nation's trash heaps. Nowhere is this truer than in the world of cosmetics and personal care products, where the lifetime of some products is a few hours or days. The lifetime of the materials from which they're made, however, can be counted in decades—or centuries.

Just look at a few of the disposables Americans throw away each year: 18 billion disposable diapers, 2 billion disposable razors, 2.2 billion disposable ballpoint pens, 2.5 billion disposable batteries, 2 million cameras, and untold billions of disposable cigarette lighters, tampon applicators, flashlights, and many other items. Indeed, according to the Coalition for Recyclable Waste, MGM/US Home Video is planning to test a disposable movie videocassette that can be viewed somewhere between 15 and 60 times before it self-destructs. After the prescribed number of playings, the tape becomes blank; you cannot re-record it, you just throw it away. Talk about planned obsolescence!

And that's just the beginning. When it comes to cosmetics and other personal care products, the list of disposables goes on and on. Almost all of this disposability, of course, is made possible by durable, enduring plastics. (See "The Perils of Plastic" for more on the lasting effects of this material.) Nearly all of these materials are neither reusable nor recyclable.

Advertising and marketing gurus point to the busy lifestyles of Americans when defending the throw-away mentality they have inspired and promoted. They tell us that they are giving us what we want: convenience. They explain that we simply don't have time to be bothered with refilling, reusing, or recycling the products we buy; we'd much rather leave these nasty details to others.

Perhaps. But what about the "nasty details" of the environment? Marketers' perceptions notwithstanding, studies indicate that most of us *do* want to be bothered with these matters, that we don't want to knowingly contribute to the earth's problems. Maybe, just maybe, the conveniences offered by our throw-away society are as temporary as the products themselves.

## Diapers, Diapers, and More Diapers

The number of disposable diapers we use and throw away each day is simply staggering: more than 18 billion diapers, containing an estimated 2.8 million tons of excrement and urine, are dumped each year into America's dwindling landfills. That amounts to almost 44 million diapers a day, or more than 500 every *second*. The typical baby soils some 7,500 disposable diapers before graduating to the toilet. Like almost everything else that ends up in landfills, the diapers' impact on public health and the environment is unknown.

What is known is this: Disposable diapers consume some 67,500 tons of polyethylene resin each year, which is used in the waterproof plastic liners of most commercial brands. When disposed of, that plastic takes between 300 and 500 years to break down. (Or so researchers say. No one really knows; these diapers have only been around for twenty years or so.) The figures do not include the other synthetic components of most disposable diapers, including the diapers' elastic, tape tabs, adhesive, and frontal tape. Disposable diapers represent just under 2 percent of all household solid wastes—the largest single product in the waste stream after newspapers (6.8 percent) and beverage containers (5.5 percent). All told, American babies produce an estimated 3.6 million tons of used diapers a year.

That's just the beginning of the potential environmental problems. Even though parents are instructed to flush the soiled inner lining of the diaper down the toilet, one study found that fewer than 5 percent do so. An infant's feces and urine can contain any of over 100 viruses, including live polio and hepatitis from vaccine residues. These viruses are potentially hazardous to sanitation workers and to others, both through groundwater or when carried by flies.

When first introduced in the late 1950s (one of the first brands was appropriately named "Chux"), disposable diapers were not intended for everyday use. Initially, they were promoted and used largely as an alternative convenience—for families on summer vacation, for example—not as a permanent replacement for cloth diapers. As manufacturing technology improved (the first brands were as rough as sandpaper), disposable diapers gradually began to overtake cloth diapers as the swaddling of choice. Today, about 85 percent of the diaper business consists of disposables. Two companies, **Procter & Gamble** (which makes Pampers) and **Kimberly-Clark** (which makes Huggies), control a whopping 80 percent of the disposable diaper market.

**Cloth versus Disposables.** Diapers are unique among disposable plastic products: They are one of the few products in which the disposable product, besides contributing to a host of environmental problems, costs more to use than a reusable product—in this case cotton diapers from a delivery service. On average, according to a study conducted by Energy Answers Corporation, a waste management consulting firm, cotton diapers cost about 15 cents each from a delivery service, compared with 22 cents for each disposable diaper. That seven-cent difference may not seem like much, but when you consider that babies require between 6,000 and 10,000 diapers over a three-year period, the difference amounts to between $420 and $700. With cloth diapers, human waste is properly channeled into the sewage system and the fabric can be reused up to 200 times before the fabric is recycled. Almost all worn-out diaper service diapers are recycled into rags for industrial use.

One barrier to using cloth diapers is day care centers. Many require parents to provide a supply of disposable diapers for their children. But a growing number of day care centers are rediscovering diaper services, too, and are including the cost of the service in

## *Biodegradables versus Nonbiodegradables*

Here is a rough breakdown of biodegradable versus nonbiodegradable material contained in two brands of disposable diapers: Nappies and Pampers. The data were provided by Eco-Matrix, U.S. distributor of Nappies.

| | *NAPPIES* | | *PAMPERS* | |
|---|---|---|---|---|
| **Material** | **Bio.** | **Nonbio.** | **Bio.** | **Nonbio.** |
| Wood pulp | 58.8 | 0.0 | 58.2 | 0.0 |
| Nonwoven | 0.0 | 6.3 | 0.0 | 9.3 |
| Poly | 12.3 | 0.0 | 0.0 | 9.3 |
| Tissue | 7.9 | 0.0 | 7.5 | 0.0 |
| Elastic | 0.0 | 0.22 | 0.0 | 1.1 |
| Tape tabs | 0.0 | 1.08 | 0.0 | 1.6 |
| Adhesive | 4.6 | 0.0 | 0.0 | 4.5 |
| Superabsorbent | 8.8 | 0.0 | 6.6 | 0.0 |
| Waistband | 0.0 | 0.0 | 0.0 | 0.7 |
| Frontal tape | 0.0 | 0.0 | 0.0 | 1.2 |
| **TOTAL** | **92.4** | **7.6** | **72.3** | **27.7** |

parents' bills. **Kindercare**, a national chain of day-care centers, allows parents to provide cloth diapers if they submit a doctor's note stating that their child requires cloth diapers.

There have been some innovations that make the traditional washable cloth diaper even more convenient, comfortable, and durable. Contact these companies for their latest catalogs:

❑ **Baby Bunz & Co.** (P.O. Box 1717, Sebastopol, CA 95473; 707-829-5347) offers a wide range of "natural diapering" products, including Nikkys—breathable, natural fiber diapers that come in either cotton or wool felt.

❑ **Biobottoms** (Box 6009, 3820 Bodega Ave., Petaluma, CA 94953; 707-778-7945) makes Velcro-fastened, fitted wool diaper covers that help keep babies extra dry. The breathable covers can go through about five diaper changes before needing washing.

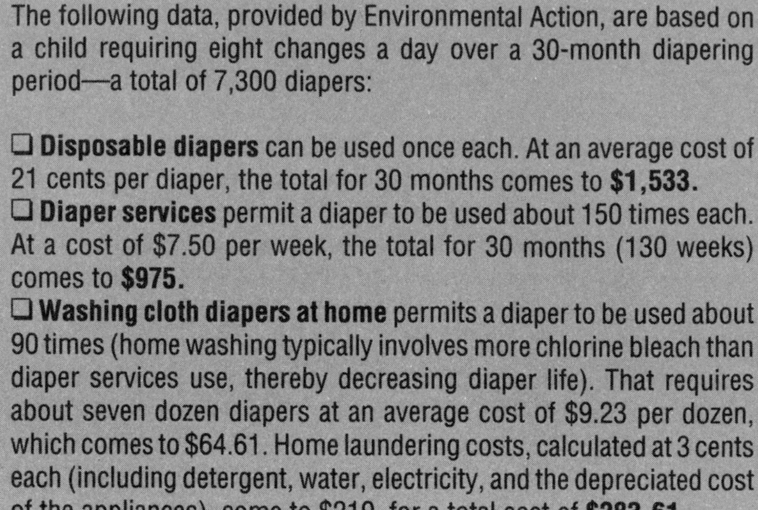

## *The Bottom Line*

The following data, provided by Environmental Action, are based on a child requiring eight changes a day over a 30-month diapering period—a total of 7,300 diapers:

❑ **Disposable diapers** can be used once each. At an average cost of 21 cents per diaper, the total for 30 months comes to **$1,533.**
❑ **Diaper services** permit a diaper to be used about 150 times each. At a cost of $7.50 per week, the total for 30 months (130 weeks) comes to **$975.**
❑ **Washing cloth diapers at home** permits a diaper to be used about 90 times (home washing typically involves more chlorine bleach than diaper services use, thereby decreasing diaper life). That requires about seven dozen diapers at an average cost of $9.23 per dozen, which comes to $64.61. Home laundering costs, calculated at 3 cents each (including detergent, water, electricity, and the depreciated cost of the appliances), come to $219, for a total cost of **$283.61.**

❑ **Bumkins** (7720 E. Redfield Rd., Suite 4, Scottsdale, AZ 85260; 606-483-7070) makes a one-piece diaper that will last through 200 washings. It features a waterproof nylon cover and absorbent thick cotton padding.

Another alternative to buying cloth diapers is to use a diaper service. They are listed in the Yellow Pages under "Diaper Services." If you cannot locate one easily, contact the **National Association of Diaper Services,** 2017 Walnut St., Philadelphia, PA 19103; 215-569-3650.

**Biodegradable Disposables.** Another alternative is a new breed of disposable diapers that contain plastics bonded to vegetable oil and to a cornstarch derivative. This material is said to break down in only two to seven years, depending on environmental conditions. (Again, these figures are entirely theoretical; little research has been done to determine these materials' biodegradability in real-life settings.) According to the manufacturers of these products, these diapers contain between 2 and 8 percent materials that are nonbiodegradable, compared with about 27 percent for other disposable diaper brands.

There are two brands of these "biodegradable" diapers available in the United States:

❑ **Nappies** (distributed by Eco-Matrix, 124 Harvard St., Boston, MA 02146; 617-730-8450)
❑ **Tender Care** (Rocky Mountain Medical Corporation, 5555 E. 71st St., Suite 8300, Tulsa, OK 74136; 918-491-9140, 800-344-6379)

Both brands are available by mail order and are available in selected stores. Prices for both are comparable with other commercial brands: about 18 to 28 cents each.

Although biodegradable diapers do not resolve all environmental issues surrounding disposable diapers—they still offer little in the way of real waste reduction—they may strike a reasonable compromise between other commercial brands and the relative inconvenience of cloth diapers. But the bottom line is that cloth diapers remain the most ecologically responsible solution—as well as the most economical.

## OTHER DISPOSABLES

Here are some other matters of concern to Green Consumers regarding disposable personal care products:

**Razors:** Americans go through about two billion disposable razors each year—the kind that include a disposable plastic handle and blade in one. Clearly, these are a waste of resources (and money). These aren't recyclable; there's no known way to separate the metal from the plastic for reuse. Because of the ready availability of reusable handles (even if they have disposable blades), the disposable variety should be avoided.

**Tampon applicators:** They have come to be called "New Jersey seashells": those plastic applicator tubes that come with some brands of tampons. When disposed of these seemingly harmless

### *Aerosol Pollution*

The environmental problems associated with aerosol cans did not end in the 1970s, when the federal government banned the use of ozone-destroying chlorofluorocarbons. The CFC ban led manufacturers of deodorants, hair sprays, and other aerosol products to substitute CFCs with chemicals that do not damage the ozone layer. Unfortunately, many of the substitutes contribute to the hydrocarbons (mostly propane, butane, and iso-butane, which are not active ingredients but help force the product out of the can) that help create smog in many urban areas. Like hydrocarbons from a car's tailpipe, they combine with nitrogen oxide emissions and sunlight to form smog.

The problem got so bad in California that in 1989 the state's Air Resources Board moved to require makers of deodorants and antiperspirants to use more environmentally acceptable substitutes. But such alternative ingredients notwithstanding, the best alternatives are products that avoid aerosol application entirely—such as pump sprays and roll-ons—which do not pollute the air.

devices often end up in rivers and streams, and, eventually, oceans. There they are mistaken for food by birds and fish. When they are consumed, they often result in these creatures choking to death. Of the three leading brands of tampons, **Tampax** offers a biodegradable cardboard applicator; **Playtex** and **Kotex** applicators are of the environmentally harmful plastic variety.

## Cruelty-Free Products

The idea of "cruelty-free" personal care products is a relatively new one, and a confusing one. On the one hand, today's consumers want products that are safe to use, that won't cause irritations or other harmful reactions. On the other hand, there is growing concern over the treatment of laboratory animals used in testing these products.

Millions of laboratory animals—mice, rats, dogs, monkeys,

and others—sacrifice their health, and often their lives, in the name of "science," testing a wide range of cosmetics and personal care products. Before it reaches your supermarket or drugstore shelves, every "new" or "improved" product goes through a battery of animal tests to make sure they are safe for human use. In the process, some 14 million animals die each year. All told, about 70 million animals are put through some kind of testing, many of which result in their painful disfigurement.

The nature of many of these tests is brutal. Test animals are routinely burned, injected with poisonous substances, artificially stressed, infected with disease, and administered electric shocks. In most cases, the tests go on for days or weeks; animals are rarely administered painkillers. One notorious experiment, the Draize Eye-Irritancy Test used in the cosmetic industry to measure the irritancy of potential new products, involves putting albino rabbits in restraining devices, then administering a few drops of the test substance into their eyes. The all-too-frequent results: the animals' eyes swell and redden until they go blind. There is also a Draize Skin-Irritancy Test. Still another test, the LD-50 (short for "lethal dose, 50 percent"), involves force feeding animals with chemicals to determine how much is required to kill half of the test animals. This test alone kills 4 million animals a year, according to the **Humane Society of the United States** (2100 L St. NW, Washington, DC 20037; 202-452-1100), one of the leading animal rights groups . Before they die, animals generally experience bleeding from the eyes, nose, and mouth; an inability to breathe; vomiting; convulsions; and paralysis.

Unfortunately, not all of these animals die for the cause of vital new products such as new cancer drugs, AIDS cures, or even baldness cures. Millions of these animals die for the cause of beauty—to create such things as mascara, shampoo, mouthwash, lipstick, hand lotion, face cream, and perfume. Even worse, according to the Humane Society, such experiments yield little to no useful information about these potential products' potential health risks to humans. Moreover, say animal protection activists, alternatives are available that require no animal testing at all, save time and money, and provide better information to protect human health.

## Cruelty-Free Companies

Despite increasingly loud protests from animal rights groups and concerned citizens, most of the leading cosmetics companies continue to use animal testing. But a growing number of small- and mid-sized cosmetics companies do not. These cruelty-free companies either rely on methods that test animals in humane ways, or do not use animal tests at all. When you purchase products from these companies, you help prevent needless animal suffering.

Keep in mind that all of these companies and products do not necessarily meet other standards important to Green Consumers. For example, many of these products—cosmetics in particular—are overpackaged, usually in plastic, creating toxic, nonbiodegradable trash. Many of these firms, however, use natural products, often minimizing allergic and other reactions in users. From the perspective of health and humanity, chances are that cruelty-free products will be superior to those from companies that continue to conduct animal tests.

The lists beginning on the nex page were compiled by **People for the Ethical Treatment of Animals** (PETA, P.O. Box 42516, Washington, DC 20015; 202-726-0156), a leading activist animal rights group. The lists are divided into the companies do not conduct animal tests, including those that do not use any animal-derived ingredients, and those that continue to conduct animal tests. To receive a catalog from any of these companies, send them $1 and a self-addressed, stamped envelope.

## *Companies That Do Not Test on Animals*

An asterisk (*) indicates companies that also do not use any animal ingredients in product formulas.

**AB Enterprises**
285 St. Mark's Place, Suite 4F
Staten Island, NY 10301

*** Abracadabra, Inc.**
P.O. Box 1040
Guerneville, CA 95446

*** Aditi Nutri-Sentials, Inc.**
P.O. Box 155
New York, NY 10012

**Advance Design Laboratories**
Box 55016, Metro Station
Los Angeles, CA 90055

*** AFM Enterprises**
1440 Stacy Court
Riverside, CA 92507

**African Bio-Botanica, Inc.**
7509 NW 13th Blvd.
Gainesville, FL 32606

**Alba Botanica Cosmetics**
1448 12th St.
Santa Monica, CA 90401

**Alexandra Avery Purely Natural**
Northrup Creek
Clatskanie, OR 97016

**Alfin Fragrances, Inc.**
15 Maple St.
Norwood, NJ 07648

*** Allen's Naturally**
P.O. Box 514
Farmington, MI 48332

**Alva-Amco Pharacal Cosmetics**
6625 Avondale Ave.
Chicago, IL 60631

**Alvin Last, Inc.**
145 Palisades St.
P.O. Box 24
Dobbs Ferry, NY 10522

**Amberwood**
Route 1
Box 206
Milner, GA 30257

*** American Merfluan, Inc.**
2479 Edison Way
Menlo Park, CA 94025

*** Ananda Country Products**
14618 Tyler Fort Rd.
Nevada City, CA 95959

**Andalina**
Tory Hill
Warner, NH 03278

**Arbonne International, Inc.**
22541 Aspan Dr.
Lake Forest, CA 92630

*** Aroma Vera Co.**
P.O. Box 3609
Culver City, CA 90231

**Atta Lavi**
443 Oakhurst Dr., #305
Beverly Hills, CA 90212

**Aubrey Organics**
4419 N. Manhattan Ave.
Tampa, FL 33614

*** Aura Cacia, Inc.**
P.O. Box 3157
Santa Rosa, CA 95402

*** Auromere Ayurvedic Imports**
1291 Weber St.
Pomona, CA 91768

**Autumn-Harp, Inc.**
28 Rockydale Rd.
Bristol, VT 05443

**Aveda Corp.**
321 Lincoln St. NE
Minneapolis, MN 55413

**Ayagutaq**
P.O. Box 176
Ben Lomond, CA 95005

*** Aztec Secret**
P.O. Box 19735
Las Vegas, NV 89132

*** Baby Touch Ltd.**
100 Sandpiper Circle
Corte Madera, CA 94925

**Bare Essentials**
809 University Ave.
Los Gatos, CA 95030

**Barsamian's**
1030 Massachusetts Ave.
Cambridge, MA 02138

**Basically Natural**
109 E. G St.
Brunswick, MD 21716

*** Baubiologie Hardware**
207-B 16th St.
Pacific Grove, CA 93950

**Baudelaire, Inc.**
Forest Rd.
Marlow, NH 03456

**Beauty Naturally**
57 Bosque Rd.
Fairfax, CA 94930

**Beauty Without Cruelty**
451 Queen Anne Rd.
Teaneck, NJ 07666

**Beehive Botanicals**
Route 8, Box 8258
Hayward, WI 54843

**Benetton Cosmetics Corp.**
540 Madison Ave, 8th Floor
New York, NY 10022

**Bevan**
P.O. Box 20072
New York, NY 10017

**Biddeford Industries, Inc.**
P.O. Box 408
Biddeford, ME 04005

*** Biogime International, Inc.**
1187 Brittore Rd.
Houston, TX 77043

*** Biokosma**
841 S. Main St.
Spring Valley, NY 10977

**Bio Line, Inc.**
8337 Pennsylvania Ave. S.
Minneapolis, MN 55431

**C. W. Bodkins Ltd.**
228 2nd Ave. SW
Pacific, WA 98047

*** Body Love**
P.O. Box 7542
Santa Cruz, CA 95061

**The Body Shop**
1341 7th St.
Berkeley, CA 94710

**The Body Shop, Inc.**
45 Horsehill Rd.
Cedar Knolls, NJ 07927

**Borlind of Germany**
P.O. Box 1487
New London, NH 03257

**Bonne Sante**
462 62nd St.
Brooklyn, NY 11220

*** Botanee**
705 Hopkins Rd.
Haddonfield, NJ 08033

**Botanicus Retail, Inc.**
7920 Greenair Dr.
Gaithersburg, MD 20879

*** Dr. E. H. Bronner**
P.O. Box 28
Escondido, CA 92025

**Bronson Pharmaceuticals**
4526 Rinetti Lane
La Canada, CA 91011

*** Bug-Off**
Route 3, Box 27A
Lexington, CA 24450

**C & S Laboratories**
5600 G McLeod Ave.
Albuquerque, NM 87109

*** Camilla Hepper**
4338 Center Gate
San Antonio, TX 78217

*** Campana Corp.**
Batavia, IL 60510

**Carbona Products Co.**
330 Calyer St.
Brooklyn, NY 11222

**Carlson Labs**
15 College Dr.
Arlington Heights, IL 60004

**Carma Laboratories**
5801 W. Airways Ave.
Franklin, WI 53132

**Carole's Cosmetics**
3081 Klondike Ave.
Costa Mesa, CA 92626

**Caswell-Massey**
111 Eighth Ave.
New York, NY 10011

*** Cernitin America, Inc.**
130 Clarkson Ave., Suite 4F
Brooklyn, NY 11226

**Certan-Dri**
Box 24845
Jacksonville, FL 32241

**Chenti Products, Inc.**
21093 Forbes Ave.
Hayward, CA 94545

**A Clear Alternative**
8707 West Lane
Magnolia, TX 77355

*** Clearly Natural Products**
P.O. Box 750024
Petaluma, CA 94975

**Clientele**
5207 NW 163rd St.
Miami, FL 33014

**Colour Quest**
616 S. 3rd St.
St. Charles, IL 60174

**Columbia Cosmetics Mfg., Inc.**
1661 Timothy Dr.
San Leandro, CA 94577

**Columbia Manicure Mfg. Co.**
1 Seneca Pl.
Greenwich, CT 06830

**Come to Your Senses**
321 Cedar Ave. S.
Minneapolis, MN 55454

*** Comfort Mfg. Co.**
1056 W. Van Buren St.
Chicago, IL 60607

*** Community Soap Factory**
P.O. Box 32057
Washington, DC 20007

**The Compassionate Consumer**
P.O. Box 27
Jericho, NY 11753

**Compassionate Products**
P.O. Box 471
Dauphin, PA 17018

**Cosmetics To Go**
29 High St.
Poole, Dorset BH15 1AB, England

**Country Comfort**
28537 Nuevo Valley Dr., Box 3
Nuevo, CA 92367

**Creature Care**
9009 South St.
Monte Rio, CA 95462

**Dermatone Laboratories, Inc.**
47 Mountain Rd.
P.O. Box 633
Suffield, CT 06078

**Desert Essence**
P.O. Box 588
Topanga, CA 90290

**Duncan Enterprises**
5673 E. Shields Ave.
Fresno, CA 93727

**Earth Science, Inc.**
P.O. Box 1925
Corona, CA 91718

**Ecco Bella**
125 Pompton Plains Crossroads
Wayne, NJ 07470

**Eco Safe Laboratories, Inc.**
Box 8702
Oakland, CA 94662

**Eco Ver Products**
c/o Mercantile Food Co.
P.O. Box 1140
Georgetown, CT 06829

**Espree**
P.O. Box 160249
Irving, TX 75016

*** Essentia**
462 62nd St.
Brooklyn, NY 11220

**EvaJon Cosmetics**
1016 E. California St.
Gainesville, TX 76240

**Fashion Two Twenty, Inc.**
1263 S. Chillicothe Rd.
Aurora, OH 44202

**Finelle Cosmetics**
137 Marston St.
Lawrence, MA 01842

**Fleur de Sante, Inc.**
P.O. Box 16090
Fort Lauderdale, FL 33318

**Flora Distributors, Inc.**
P.O. Box 67333
Vancouver, BC VSW 371, Canada

*** Forever New**
2922 West Maple
Sioux Falls, SD 57107

*** 4-D Hobe Marketing**
201 S. McKemy
Chandler, AZ 85226

*** IV Trail Products**
2504 Arthur Ave.
Edlersburg, MD 21784

**Freeman Cosmetics Corp.**
P.O. Box 17
Hollywood, CA 90078

**A. J. Funk & Co.**
1471 Timber Dr.
Elgin, IL 60120

**Future Perfect**
1341 Seventh St.
Berkeley, CA 94710

**G. T. International**
1800 S. Robertson Blvd., Suite 182
Los Angeles, CA 90035

**General Nutrition**
1301 39th St.
Fargo, ND 58107

**Giovanni Cosmetics, Inc.**
3023 N. Collidge Ave.
Los Angeles, CA 90039

**Going First Class**
P.O. Box 266
Pocono Manor, PA 18349

*** Golden Lotus, Inc.**
Box 40189
Grand Junction, CO 81504

**Golden Pride/Rawleigh**
1501 Northpoint Parkway
West Palm Beach, FL 33407

**Golden Star, Inc.**
P.O. Box 12539
N. Kansas City, MO 64116

**Goldwell Cosmetics**
9015 Junction Dr.
Annapolis Junction, MD 20701

*** Granny's Old-Fashioned Products**
3581 E. Milton St.
Pasadena, CA 91107

**Gruene, Inc.**
1621 W. Washington Blvd.
Venice, CA 90291

**Hain Pure Food Co.**
13660 S. Figueroa
Los Angeles, CA 90061

**Dr. Hauschka Cosmetics, Inc.**
Meadowbrook West
Wyoming, RI 02898

**Hawaiian Resources**
1123 Kapahulu Ave.
Honolulu, HI 96816

**Heart's Desire**
1307 Dwight Way
Berkeley, CA 94702

**Heavenly Soap**
5948 E. 30th St.
Tucson, AZ 85711

**Home Health Products, Inc.**
P.O. Box 3130
Virginia Beach, VA 23454

*** Home Service Products Co.**
230 Willow St.
Bound Brook, NJ 08805

*** Huish Chemical Co.**
3540 W. 1987 S.
Salt Lake City, UT 84125

**Humane Alternative Products**
8 Hutchins St.
Concord, NH 03301

**Humane Street USA**
467 Saratoga Ave., Suite 300
San Jose, CA 95129

**Humphreys Pharmacal, Inc.**
63 Meadow Rd.
Rutherford, NJ 07070

**I Care Cosmetics**
1937 Kenwood St.
Burbank, CA 91505

**Ida Grae Products**
424 Laverne Ave.
Mill Valley, CA 94941

**Image Laboratories**
721 S. San Pedro
Los Angeles, CA 90014

**Indra**
321 Lincoln St. NE
Minneapolis, MN 55413

**Institute of Trichology**
1619 Reed St.
Lakewood, CO 80215

*** Integrated Health, Inc.**
1661 Lincoln Blvd., Suite 300
Santa Monica, CA 90404

*** International Rotex, Inc.**
P.O. Box 20697
Reno, NV 89515

**International Vitamin Corp.**
P.O. Box 1746
Union, NJ 07083

**InterNatural**
P.O. Box 680
South Sutton, NH 03273

**Internatural Distributors**
RFD Baker Hill Rd.
Bradford, NH 03221

**Intl. Oriental Beauty Secrets**
1800 S. Robertson Blvd., #182
Los Angeles, CA 90035

**Irma Shorell, Inc.**
720 Fifth Ave.
New York, NY 10019

**Jacki's Magic Lotion**
258 A St., #7A
Ashland, OR 97520

**James Austin Co.**
P.O. Box 827
Mars, PA 16046

**Jason Natural Products**
8468 Warner Dr.
Culver City, CA 90232

**John F. Amico & Co.**
7327 W. 90th St.
Bridgeview, IL 60455

**John Paul Mitchell Systems**
20801 Nordhoff St.
Chatsworth, CA 91311

## *The Body Shop*

Perhaps no company better exemplifies the green entrepreneurial spirit better than The Body Shop. The British-based company, which now has several shops and a mail-order business in the U.S., has achieved great success by marketing a high-quality product with the environment in mind.

Founded in 1976, the firm focused on the growing segment of the marketplace interested in products that were healthy for both individuals and the planet. In its shops and by mail, The Body Shop sells products made from natural ingredients—with minimal packaging, including biodegradable plastics—whose ingredients have not been tested on animals during the past five years. Moreover, the company has sought to work with other individuals and organizations, including Friends of the Earth and Greenpeace, to use its shops as a forum on a range of environmental and social issues. Posters on subjects such as acid rain, endangered species, and ozone depletion appear in store windows, and handouts on these subjects are available in stores as well as from the company's catalog.

Not all of the company's trademark environmental practices have been implemented in the United States. For example, in its British stores, the company offers a bottle refill service: customers bring in their empty product bottles, which are then refilled and returned, thereby saving packaging costs for the second and subsequent purchase of the same product. In the U.S., however, restrictive sanitation laws prohibit this practice. Still, the U.S. stores will accept empty product containers, which they are attempting to recycle.

More recently, The Body Shop has begun a campaign to create products made from rain forest herbs, in an attempt to demonstrate sustainable agriculture in the Amazonian rain forests. These products will be available in the coming months.

For more information about The Body Shop, or to obtain a catalog, contact them at 45 Horsehill Rd., Hanover Technical Center, Cedar Knolls, NJ 07927; 201-984-2535, 800-541-2535.

**Jurlique Cosmetics**
16 Starlit Dr.
Northport, NY 11768

**Kallima**
915 Whitmore Dr.
Rockwall, TX 75087

**Kimberly Sayer, Inc.**
61 W. 82nd St., #5A
New York, NY 10024

**Kiss My Face**
P.O. Box 804
New Paltz, NY 12561

**KMS Research, Inc.**
4712 Mountain Lakes Blvd.
Redding, CA 96003

*** KSA Jojoba**
19025 Parthenia St., #200
Northridge, CA 91324

*** LaCrista, Inc.**
P.O. Box 240
Davidsonville, MD 21035

*** L'anza Research Laboratories**
5523 Ayon Ave.
Irwindale, CA 91706

**Levlad, Inc.**
9183-5 Kelvin St.
Chatsworth, CA 91311

**Life Tree Products**
1448 12th St.
Santa Monica, CA 90401

**Lion & Lamb**
29-28 41st Ave.
Long Island City, NY 11101

**Livos Plant Chemistry**
614 Agua Fria St.
Santa Fe, NM 87501

**Loanda Herbal Products**
84 Galli Dr.
Novato, CA 94947

*** Lumen Food Corp.**
2116 Hodges
Lake Charles, LA 70601

*** M&N Natural Products**
P.O. Box 4502
Anaheim, CA 92803

**McGean-Rohco, Inc.**
2910 Harvard Ave.
Cleveland, OH 44109

**The Magic of Aloe, Inc.**
7300 N. Crescent Blvd.
Pennsauken, NJ 08110

**Marie Lacoste Enterprises**
1059 Alameda de las Pulgas
Belmont, CA 94002

**Martha Hill Cosmetics**
5 Ivy Court
Metuchen, NJ 08840

*** Martin Von Myering**
422 Jay St.
Pittsburgh, PA 15212

**Marly Savon Clair**
P.O. Box 54841
Terminal Annex
Los Angeles, CA 90054

**Mercantile Food Co.**
4 Old Mill Rd.
Georgetown, CT 06829

## *Companies That Do Test on Animals*

Alberto-Culver Co.
American Cyanamid Co.
Andrea Rabb
Aramis, Inc.
Armour-Dial, Inc.
BeautiControl Cosmetics
Beecham Cosmetics, Inc.
Bonne Belle, Inc.
Boyle-Midway
Breck
Bristol-Myers Co.
Carter Wallace, Inc.
Chanel, Inc.
Chesebrough-Ponds, Inc.
Church & Dwight
Clairol, Inc.
Clarins of Paris
Clinique Laboratories, Inc.
Clorox Co.
Colgate-Palmolive Co.
Cosmair, Inc.
Coty
Dana Perfumes Corp.
Del Laboratories
Diversey Wyandotte Corp.
Dorothy Gray
Dow Chemical Co.
Drackett Products Co.
Economics Laboratory, Inc.
Eli Lilly & Co.
Estée Lauder, Inc.
Frances Denny
Gillette Co.
Givaudan Corp.
Helena Rubenstein
Helene Curtis Industries, Inc.
Houbigant, Inc.
Jean Patou, Inc.
Jergens
Johnson & Johnson
S.C. Johnson & Son, Inc.
Johnson Products Co., Inc.
Jovan, Inc.
Lamaur, Inc.
Lancome
Lever Brothers, Inc.
L'Oreal
Maybelline
Mennen Co.
Neutrogena
Nina Ricci
Pfizer
Physicians Formula Cosmetics
Procter & Gamble Co.
Purex Corp.
Quintessence, Inc.
Richardson-Vicks, Inc.
Schering-Plough (Maybelline)
Sea & Ski Corp.
Shulton
Squibb
Sterling Drug, Inc.
Syntex
Texize
Vidal Sassoon, Inc.
Warner-Lambert Co.
Wella Corp.
Westwood Pharmaceuticals
Zotos International, Inc.

*Courtesy PETA*

*** Mia Rose Products, Inc.**
1374 Logan Ave., Unit C
Costa Mesa, CA 92626

**Michael's Health Products**
7040 Alamo Downs Pkwy.
San Antonio, TX 78238

*** Microbalanced Products**
25 Aladdin Ave.
Dumont, NJ 07628

**Mild and Natural**
84 Galli Dr.
Novato, CA 94947

**Miss C's Closet**
524A Bloomfield Ave.
Verona, NJ 07044

*** Mountain Fresh Products (Golden Lotus)**
Box 40189
Grand Junction, CO 81504

**Mountain Herbery**
84 Galli Dr.
Novato, CA 94947

**Mountain Ocean Ltd.**
1738 Pearl
Boulder, CO 80302

**The Murphy-Phoenix Co.**
P.O. Box 22930
Beachwood, OH 44122

**My Brother's Keeper, Inc.**
P.O. Box 1769
Richmond, IN 47375

**Naturade Cosmetics**
7100 E. Jackson St.
Paramount, CA 90723

**Natural Organics, Inc.**
10 Daniel St.
Farmingdale, NY 11735

**Nature Basics**
61 Main St.
Lancaster, NH 03584

**Nature Cosmetics, Inc.**
881 Alma Real, Suite 101
Pacific Palisades, CA 90272

**Nature de France**
444 Park Ave. S.
New York, NY 10016

*** Nature's Plus**
10 Daniel St.
Farmingdale, NY 11735

**Nature's Gate Herbal Cosmetics**
9183-5 Kelvin St.
Chatsworth, CA 91311

**Naturessence**
881 Alma Real, Suite 101
Pacific Palisades, CA 90272

**Nelson Chemicals Co.**
12345 Schaefer Hwy.
Detroit, MI 48227

*** New Age Products**
16100 N. Highway 101
Willits, CA 95490

*** Neway**
Little Harbor
Marblehead, MA 01945

*** New World Minerals**
4459 E. Rochelle Ave.
Las Vegas, NV 89121

**Nexxus**
P.O. Box 1274
Santa Barbara, CA 93116

*** No Common Scents**
King's Yard
220 Xenia Ave.
Yellow Springs, OH 45387

**North Country Soap**
7888 County Rd., #6
Maple Plain, MN 55359

**Nu Skin Intl., Inc.**
145 East Center
Provo, UT 84601

**Nutri-Metics Intl., Inc.**
19501 E. Walnut Dr.
City of Industry, CA 91748

**O'Naturel, Inc.**
535 Cordova Rd., #472
Santa Fe, NM 87501

**Oriflame International**
76 Treble Cove Rd.
North Billerica, MA 01862

**Orjene Natural Cosmetics**
5-43 48th Ave.
Long Island City, NY 11101

**Painlessly Beautiful**
1260 Lumber St.
Middletown, PA 17057

**Panacea II**
P.O. Box 294
Columbia, PA 17512

**Patricia Allison**
4470 Monahan Rd.
La Mesa, CA 92041

**Paul Penders USA**
1340 Commerce St.
Petaluma, CA 94954

**The Peaceable Kingdom**
1902 W. 6th St.
Wilmington, DE 19805

**PetGuard**
165 Industrial Loop S., Unit 5
Orange Park, FL 32073

*** Pets 'n' People, Inc.**
5312 Ironwood St.
Rancho Palos Verdes, CA 90274

**Premier Industrial Corp.**
4500 Euclid Ave.
Cleveland, OH 44103

**Professional & Technical Services**
3333 N.E. Sandy Blvd., #208
Portland, OR 97232

**Puritan's Pride**
105 Orville Dr.
Bohemia, NY 11716

**Rachel Perry, Inc.**
9111 Mason Ave.
Chatsworth, CA 91311

**Rainbow Concepts**
P.O. Box 2332
Stone Mountain, GA 30086

**Rainbow Research Corp.**
170 Wilbur Pl.
Bohemia, NY 11716

**W.T. Rawleigh Co.**
223 E. Main St.
Freeport, IL 61032

**The Real Aloe Co.**
4735-4D Industrial St.
Simi Valley, CA 93063

**Red Saffron**
3009 16th Ave. S.
Minneapolis, MN 55407

**Reviva Labs, Inc.**
705 Hopkins Rd.
Haddonfield, NJ 08033

**Richlife, Inc.**
2211 E. Orangewood
Anaheim, CA 92806

* **I. Rokeach & Sons, Inc.**
560 Sylvan Ave.
Englewood Cliffs, NJ 07632

* **Royal Laboratories, Inc.**
465 Production St.
San Marcos, CA 92069

* **Sappo Hill Soapworks**
654 Tolman Creek Rd.
Ashland, OR 97520

**Schiff**
121 Moonachie Ave.
Moonachie, NJ 07074

**Sebastian International, Inc.**
6160 Variel Ave.
Woodland Hills, CA 91746

* **The Shahin Soap Co.**
427 Van Dyke Ave.
Haledon, NJ 07538

**Sheffield Industries**
P.O. Box 351
New London, CT 06320

**Shikai Products**
P.O. Box 2866
Santa Rosa, CA 95405

**Sierra Dawn**
P.O. Box 1203
Sebastopol, CA 95472

* **Simplers Botanical Co.**
Box 39
Forestville, CA 95436

* **Sirena Tropical Soap Co.**
P.O. Box 31673
Dallas, TX 75231

* **Sleepy Hollow Botanicals**
84 Galli Dr.
Novato, CA 94947

* **The Soap Factory**
1510 Randolph St., #205
Carrollton, TX 75006

**Solventol Chemical Products, Inc.**
13177 Huron River Dr.
Romulus, MI 48174

**Sombra Cosmetics, Inc.**
5600-G McLeod Ave.
Albuquerque, NM 87109

**Sorik International**
3116 DeSalvo Rd.
Jacksonville, FL 32216

**Spare the Animals, Inc.**
P.O. Box 233
Twerton, RI 02878

**St. Ives, Inc.**
944 Indian Peak Rd.
Rolling Hills, CA 90274

**Sunrise Lane Products, Inc.**
780 Greenwich St.
New York, NY 10014

**Sunshine Makers, Inc.**
15922 Pacific Coast Hwy.
Huntington Harbor, CA 92049

**Sunshine Scented Oils**
1919 S. Burnside Ave.
Los Angeles, CA 90016

**Tom's of Maine**
Railroad Ave.
Kennebunk, ME 04042

**Trans India Products**
P.O. Box 2866
Santa Rosa, CA 95405

**Tyra Skin Care, Inc.**
9427 Lurline Ave.
Chatsworth, CA 91311

**Van Straaten Chemical Co.**
630 W. Washington Blvd.
Chicago, IL 60606

*** Vegan Street**
P.O. Box 5525
Rockville, MD 20855

**Velvet Products Co.**
P.O. Box 5459
Beverly Hills, CA 90210

**Vita Wave Products**
7131 Owensmouth Ave., #94D
Canoga Park, CA 91303

**Vivaiane Woodard Cosmetics**
7712 Densmore Ave.
Van Nuys, CA 91406

**Wachter's Organic Sea Products**
360 Shaw Rd.
South San Francisco, CA 94080

*** Warm Earth Cosmetics**
334 W. 19th St.
Chico, CA 95928

**Watkins, Inc.**
150 Liberty St.
Winona, MN 55987

**Wite-Out**
10114 Bacon Dr.
Beltsville, MD 20705

**Weleda, Inc.**
841 S. Main St.
Spring Valley, NY 10977

*** Without Harm**
4605 Pauli Dr.
Manilus, NY 13104

**Wysong Corp.**
1880 N. Eastman Rd.
Midland, MI 48640

*** Youthessence Ltd.**
P.O. Box 3057
New York, NY 10185

**Zia Cosmetics**
950 Bay St., Suite 5
San Francisco, CA 94109

## Mail-Order Companies

In addition to The Body Shop (see box on page 233), here is a select list of mail-order companies that feature cruelty-free and environment-friendly products:

❑ **The Compassionate Consumer** (P.O. Box 27, Jericho, NY 11753; 718-445-4134) distributes a variety of cruelty-free products that don't contain animal ingredients, from perfumes to skin care products to hair care products to cosmetics.

❑ **Co-Op America** (2100 M St. NW, Suite 310, Washington, DC 20036; 202-872-5307, 800-658-5507) sells a variety of cruelty-free products, including an herbal skin care system, herbal shaving creams, and sun protection creams and sprays.

❑ **Earthen Joys** (1412 11th St., Astoria, OR 97103; 503-325-0426, 800-451-4540) sells bath products and skin cleansers.

❑ **Ecco Bella** (125 Pompton Plains Crossroads, Wayne, NJ 07470; 201-890-7077) sells cruelty-free products ranging from cosmetics, moisturizing cream, facial cleansers, and skin toners to suntan lotion, massage oils, and perfumes.

❑ **Humane Alternative Products** (8 Hutchins St., Concord, NH 03301; 603-224-1361) distributes a variety of cruelty-free products, including hair care, skin care, cosmetics, perfumes, and colognes.

❑ **Sunrise Lane** (780 Greenwich St., New York, NY 10014; 212-242-7014) sells products for baby care, mouth care, healing, sun care, skin care, hair care, and cosmetics.

❑ **Vegan Street** (P.O. Box 5525, Rockville, MD 20855; 301-869-0086) offers antiperspirants, baby and bath products, conditioners and gels, shampoos, cosmetics, hair sprays, men's products, oral care, perfumes, soaps and cleansers, and sun care products.

❑ **Walnut Acres** (Penns Creek, PA 17862; 717-837-0601) sells Aubrey Organics products, made from natural vegetable sources with no perfumes, artificial colorings, or preservatives. Products available range from soap and shampoo to deodorant and moisturizing cream.

# G·R·E·E·N T·R·A·V·E·L

Tourism is one of the fastest-growing industries in the world. Unfortunately, an increasing number of travelers are visiting places that are economically oppressed and that have few restrictions on tourism and little protection for the natural environments that often lure tourists in the first place.

Our rush to reach distant and exotic destinations has created many problems. Trekkers in Nepal are destroying land terraces that grow fodder for cattle. To build airports capable of accommodating jumbo jets, large areas of forests are being destroyed in Indonesia. In the Caribbean, the number of tourists is often larger than the local populace. Remote villages are being razed to build access highways to new luxury resorts. Even in the United States, many travelers looking for a "back to nature" experience flock to national parks and forests, crowding out the natural wildlife, trampling flora and fauna, and leaving trails of waste and garbage. And even travel for the masses has its environmental impact. Huge cruise ships sail through pristine waters, for example, dumping their wastes along the way.

Touring and the environment can peacefully coexist. But there are some general guidelines to avoid disrupting the local culture or environment. Wherever you travel, you should be sensitive to and respect the native culture. You should not disturb wildlife and you should dispose of waste in the same way you would at home: in some appropriate receptacle.

As the need for responsible travel has become increasingly

apparent, several organizations have drafted model guidelines for what has come to be called "ecotourism," but which might equally be dubbed "green travel" or "low-impact travel." The last term, in fact, may be most descriptive: Your goal as a traveling Green Consumer is to have minimum impact on the areas in which you travel. To paraphrase the classic Sierra Club motto, you should leave nothing but footprints, and take nothing but photographs and memories.

There is a growing number of local and national organizations engaging in ecological or low-impact travel. The **National Audubon Society**, for example, has established an environmental code of ethics for responsible tourism and encourages all individuals and groups to adopt it. For more information contact: Audubon Travel Programs, National Audubon Society, 950 Third Ave., New York, NY 10022; 212-832-3200. The **North American Coordinating Center for Responsible Tourism** (2 Kensington Rd., San Anselmo, CA 94960; 415-258-6594) provides a "Code of Ethics for Travelers," developed by the Christian Conference of Asia, which, among other things, calls for "travel in a spirit of humility and with a genuine desire to meet and talk with the local people," and for discovering "the enrichment that comes from seeing another way of life, rather than looking for the `beach paradise' of the tourist posters."

## Environmental Travel Organizations

There are many options for selecting an environmentally sensitive and educational trip. A growing number of environmental groups are providing their own natural history and ecologically oriented expeditions to all corners of the earth. From about $300 and up for a domestic trip or $1,500 and up for an overseas journey, you can boat through the jungles of the Amazon, whale watch in Argentina, bird watch in Bali or Komodo, and visit such regions as the Galápagos Islands, India, Peru, the Seychelles, Madagascar, or New Zealand. On the domestic front there are wildlife photography trips to Alaska, journeys to observe the beauty and tranquility of the Southwest, and diving trips off the coral reefs of the Florida Keys.

Here are a few of the organizations offering such trips:

The Nature Conservancy
1250 24th St. NW
Washington, DC 20037
202-293-4800

Sierra Club
730 Polk St.
San Francisco, CA 94109
415-776-2211

World Wildlife Expeditions
1250 24th St. NW
Washington, DC 20037
202-293-4800

National Audubon Society
950 Third Ave.
New York, NY 10022
212-832-3200

Massachusetts Audubon Society
S. Great Rd.
Lincoln, MA 01773
617-259-9500, 800-289-9504

A few groups of note include **Earthwatch** (680 Mt. Auburn St., P.O. Box 403, Watertown, MA 02272; 617-926-8200), a nonprofit organization that funds scientific research expeditions. Travelers become members of research teams and share in both labor and costs of field research; part of the cost is tax deductible. Past programs include study of the environment of Australia's rain forest canopy, anthropologic digs that uncovered 4,000-year-old remains in Majorca, a study of spinner dolphins in Moorea, and observation of wildlife in the Blue Ridge Mountains of Virginia. Another organization, the **Amazon Conservation Foundation** (18328 Gulf Blvd., Indian Shores, FL 34635; 813-391-6211), organizes trips that help support conservation efforts in Peruvian jungles.

Another group, the **Earth Preservation Fund** (EPF), is a nonprofit organization that sponsors such projects as reforestation in the Himalayas and cleanup hikes along the Inca Trail. EPF has information about responsible tourism and can arrange travel through its sister organization, **Wildland Journeys** (3516 NE 155th St., Seattle, WA 98155; 206-365-0686, 800-345-4453). You may wish to contact them about future programs and trips.

## Travel Operators and Clubs

Adventure travel is big business. There are guides and outfitters willing to take you almost anywhere on earth—if you're willing to pay the price. But in their quest to make a buck, some of these operators are destroying our planet's last frontiers. There are, however, an increasing number of groups joining in the effort to treat the earth and its people in sensitive ways.

How can you tell a "green" tour operator from the others? The best way is to ask questions. Find out if the organization is ecologically sensitive, whether it provides low-impact travel, and if it follows the National Audubon Society's travel ethic. Some of the questions to ask about their trips:

- ❑ Are guides trained to be ecologically conscious and environmentally sensitive?
- ❑ Are wildlife and their habitats free from stress so you do not inhibit animal behavior?
- ❑ Are wildlife areas to be visited carefully managed?
- ❑ Which conservation efforts will this tour support?
- ❑ Will the guide(s) be able to conscientiously warn against types of souvenirs or mementos that threaten wildlife or plant life?

Here is a list of some of the organizations dedicated to green travel and a few adventure travel operators that specialize in nature tours:

**ABEC's Alaska Adventures** (1304 Westwick Dr., Fairbanks, AK 99712; 907-457-8907) offers minimum-impact wilderness backpacking and river trips through Alaska's parks and wildlife refuges.

**Above the Clouds Trekking** (P.O. Box 389, Worcester, MA 01602; 508-799-4499) features in-depth cultural experiences, including trekking in the Himalayas, Andes, and Europe.

**Adventure Center** ( 5540 College Ave., Oakland, CA 94618; 415-654-1879, 800-228-8747) offers safaris, cycling, sailing, and cultural interaction in more than 80 countries.

**Adventure Source International** (5353 Manhattan Cir., Ste. 103, Boulder, CO 80303; 303-499-2296) offers diving, river rafting, and cultural and natural history trips to Australia, Asia, and North America.

**Adventure Specialists, Inc.** (Bear Basin Ranch, Westcliff, CO 81252; 719-783-2519, 800-777-4771) features horseback trips, rafting, mountain biking, and climbing adventures in Colorado, Mexico, and South America.

**The Adventure Works USA** (P.O. Box 37, Winter Park, CO 80482; 303-726-9192, 800-274-0571) books mountain bike tours of the Arapahoe National Forest, with emphasis on the forest's geology and wildlife.

**Alaska Discovery** (369 Franklin St., Juneau, AK 99802; 907-586-1911) arranges kayak, canoe, and rafting trips; expeditions; and camping.

**American River Touring Association** (Star Route 73, Groveland, CA 94601; 209-962-7873) features white-water rafting through rivers in California, Utah, Oregon, and Wyoming's Grand Canyon National Park.

**American Youth Hostels** (P.O. Box 37613, Washington, DC 20013; 202-783-6161) offers inexpensive trips throughout the world by providing simple accommodations and organized biking, hiking, and motoring travel for members. AYH is based on the youth hostel concept that evolved in Germany to promote world peace: that by understanding each other and our environment we are better able to solve world problems.

**Appalachian Mountain Club** (P.O. Box 298, Gorham, NH 03581; 603-466-2721) offers working vacations on trail projects around the U.S.

**Asian Pacific Adventures** (336 Westminster Ave., Los Angeles, CA 90020; 213-935-3156) offers photographic, bicycling, and hiking tours, with emphasis on experiencing other cultures and natural environments.

**Bikecentennial** (P.O. Box 8308, Missoula, MT 59807; 406-721-1776) organizes North American bicycle trips.

**Biological Journeys** (1696 Ocean Dr., McKinleyville, CA 95521; 707-839-0178) features natural history marine wilderness tours in Baja California, Peru, Australia, New Zealand, and Alaska.

**Breakaway Adventure Travel** (94 Sherman St., Cambridge, MA 02140; 617-497-0855) books worldwide travel using local forms of transportation—feet, horses, rafts, etc.

**CEDAM International** (Fox Rd., Croton-on-Hudson, NY 10520; 914-271-5365) organizes volunteer expeditions to assist scientists with marine projects in the Caribbean, Pacific Ocean, and Indian Ocean.

**Cheeseman's Ecology Safaris** (20800 Kittredge Rd., Saratoga, CA 95070; 408-867-1371) features ecology safaris to East Africa, the Seychelles, Australia, New Guinea, South America, and Central America.

**Colorado Outward Bound** (945 Pennsylvania St., Denver, CO 80203; 303-837-0880) offers mountaineering, canyon exploration, and white-water rafting in Colorado, Utah, Alaska, Kansas, Texas, and some overseas destinations.

**Co-Op America's Travel Links** (2100 M St. NW, Suite 310, Washington, DC 20063; 800-992-1903) offers "responsible" tourism—in-depth cultural and educational tours to South America, Europe, Africa, and China.

**Explorers at Sea** (P.O. Box 51, Stonington, ME 04681; 207-367-2356) specializes in sea kayaking adventures and island camping.

**InnerAsia Expeditions** (2627 Lombard St., San Francisco, CA 94123; 415-922-0448) offers natural history trips to China, Tibet, Kashmir, Japan, and the Soviet Union, with focus on travel that minimally affects the environment.

**Joseph Vanos Photo Safaris** (Box 655, Vashon Island, WA 98070; 206-463-5383) provides photography and natural history tours throughout the U.S., Canada, Africa, and Central America.

**Lemur Tours** (2562 Noriega St., San Francisco, CA 94122; 415-681-8222) features nature and wildlife tours to Madagascar.

**Nature Expeditions International** (474 Willamette, P.O. Box 11496,

## *Adventures for Special People*

**Environmental Traveling Companions** (Fort Mason Ctr., Landmark Building C, San Francisco, CA 94123; 415-474-7662) specializes in wilderness adventures for people with special needs, including disadvantaged youth and people of all ages who are visually or hearing impaired, or physically, emotionally, or developmentally disabled. Trips span a wide range of activities, including sea kayaking, Nordic skiing, and white-water rafting. The group can provide school trips customized to meet the needs of teachers and students. Other individuals can participate in these trips by volunteering as guides.

Eugene, OR 97440; 503-484-6529) specializes in worldwide natural history and wildlife expeditions, focusing on photography and anthropology.

**Oceanic Society Expeditions** (Fort Mason Ctr., Bldg. E, Rm. 240, San Francisco, CA 94123; 415-441-1106) is dedicated to the protection, understanding, and management of lecan and coastal environments, and promotes responsible tourism.

**Off the Deep End Travels** (P.O. Box 7511, Jackson, WY 83001; 307-733-8707) features bicycling, kayaking, rafting, and trekking throughout Asia, the South Pacific, and Europe. The focus is on understanding cultures by living as the natives live.

**REI Adventures** (P.O. Box 8090, Berkeley, CA 94707; 415-526-4005) specializes in a variety of culturally and environmentally sensitive trips, throughout the world.

**Safaricentre International** (3201 N. Sepulveda Blvd., Manhattan Beach, CA 90266; 213-546-4411) features wildlife and ecology safari tours throughout Asia, Africa, and South America.

**Sanctuary Travel Services** (3701 E. Tudor Rd., Anchorage, AK 99507; 907-561-1212, 800-247-3149) features environmentally conscious trips throughout Alaska. The group gives 20 percent of its commissions to support environmental organizations.

**South American Explorer's Club** (P.O. Box 18327, Denver, CO 80218; 303-320-0388) is a nonprofit organization for travelers, hikers, scientists, and others interested in South America, with clubhouses in Lima, Peru, and in Quito, Ecuador. The club's information center in Denver offers specialized trip planning and information on scientific expeditions.

**Victor Emanuel Nature Tours** (P.O. Box 33008, Austin, TX 78764; 512-328-5221) offers birding and natural history tours worldwide with a commitment to conservation.

**Voyagers International** (P.O. Box 915, Ithaca, NY 14851; 607-257-3091) offers nature photography throughout Africa, Australia, Asia, and Antarctica.

**Wilderness Travel** (801 Allston Way, Berkeley, CA 94710; 415-548-0420, 800-247-6700) offers wildlife- and nature-oriented trips as an active learning experience to increase awareness of preservation worldwide. Areas of interest include the Himalyas, South America, East Africa, and Europe.

## Educational Organizations and Volunteer Programs

In addition to the offerings through environmental organizations and travel outfitters, several educational centers provide natural history tours throughout the United States. Travelers can obtain credit for volunteering or participating in environmental programs offered by some of these programs.

**Caretta Research Project—Savannah Science Museum** (4405 Paulsen St., Box B, Savannah, GA 31405; 912-355-6705) offers trips in which volunteers spend one week assisting project biologists monitor the nesting of endangered loggerhead sea turtles at the Wassaw National Wildlife Refuge and Georgia Barrier Island near Savannah. This is a volunteer conservation and research project rather than a strictly recreational travel opportunity.

**National Audubon Society Expedition Institute** (Northeast Audubon Center, Sharon, CT 06069; 203-364-0522), in conjunction with Lesley College, offers unique courses leading to B.S. and M.S. degrees in environmental education. The program combines classroom study with one to two years of expedition courses. The program also offers a high school course that allows students to fulfill one year of accredited study.

**Recursos** (826 Camino de Monte Rey, Suite A-3, Santa Fe, NM 87501; 505-982-9301) features natural history study trips investigating the delicately balanced ecologies in the desert Southwest. Travelers learn about such issues as desertification, decreasing plant and animal species, climatic changes, and the greenhouse effect.

**Society for Ecology Restoration** (University of Wisconsin, Arboretum, 1207 Seminole Highway, Madison, WI 53711; 608-263-7889) was established to promote research into the restoration, creation, and management of biotic communities. Participants provide hands-on help with ecological restoration projects in such places as Yellowstone National Park, Rocky Mountain National Park, and Afton State Park on the Saint Croix River in Minnesota.

**Smithsonian National Associate Program** (Smithsonian Institution, 1100 Jefferson Dr. SW, Rm. 3045, Washington, DC 20560; 202-357-2627) conducts natural history study tours in a variety of locations. Foreign trips include China, India, Mongolia, and the Himalayas. Domestic destinations include Death Valley, Okefenokee Swamp, Sanibel Island, Alaska, and Appalachia.

**Student Conservation Association** (P.O. Box 550, Charlestown, NH 03603; 603-826-5741) is a nonprofit educational organization providing students and others the opportunity to volunteer with resource management agencies maintaining national parks, forests, and public lands. A wide range of opportunities is available, with travel expenses and lodging provided.

**University Research Expeditions Program** (University of California, Berkeley, CA 94720; 415-642-6586) offers research expeditions emphasizing research that can be applied to improving people's lives and preserving the earth's resources. Travelers become part of a team and study such diverse subjects such as animal behavior, anthropology, sociology, archaeology, botany, and ecology in places such as Kenya, Indonesia, the United States, and the Fiji Islands.

## Guest Houses and Resorts

**Chanchich Lodge** (P.O. Box 37, Beliz City, Beliz; 800-343-8009 in the U.S., 501-2-77031 locally) offers a lodge and cabins in a tropical wilderness setting. Located in an ancient Mayan plaza, it features trails, wildlife, and Mayan ruins.

**Hawaiian Center for Ecological Living** (Star Route 13008, Keaau, HI 96749; 808-966-8592) offers a tropical homestead and guest house on the Island of Hawaii, along with alternative tours of the islands.

**Lost Valley Center** (81868 Los Valley Lane, Dexter, OR 97431; 503-937-3351) is a family retreat and conference center located at the edge of the Willamett Valley near 200 square miles of national forests. The center provides campsites, cabins, and some dorm-style bunk units in a wilderness setting with many hiking trails and creeks and rivers nearby.

**Maho Bay Camps, Inc.** (17 E. 73rd St., New York, NY 10021; 212-472-9453, 800-392-9004) offers environmentally planned camping on Saint John, U.S. Virgin Islands. The organization is also in the process of building a full-fledged resort area on Saint John, with flora and fauna restored to the pre-Columbus era, cabañas built off the ground to preserve tortoise habitat, and an emphasis on native wildlife.

## Closer to Home

Many communities are involved in preserving the local environment. Of course, there may be no better place to learn about the natural environment than a local zoo. A growing number of city zoos are developing environments beyond cement cages, emphasizing animals' natural habitats. Contact the **American Associa-**

**tion of Zoological Parks and Aquariums** (Oglebay Park, Wheeling, WV 26003; 304-242-2160) for information about zoological parks and aquariums in your area.

There are many other ways to participate in environmental travel locally. Contact local environmental groups or local travel agencies for referrals to travel outfitters and individuals in your area that are involved in responsible and environmental tourism. Below is a list of national organizations and government agencies that can provide information about local hiking, national parks, forests, seashores, and historical sites.

**American Hiking Society** (1015 31st St. NW, Washington, DC 20007; 703-385-3252) encourages hikers to build and maintain trails. Their "Volunteer Vacations" organize crews of volunteers to work on trail construction and maintenance for two-week periods on public lands across the U.S.

**U.S. Forest Service** (Office of Public Affairs, Rm. 3008, South Bldg., Washington, DC 20250; 202-447-6661) has maps and information available for each of its 122 national forests.

**National Park Service** (Office of Public Inquiry, P.O. Box 37127, Washington, DC 20013; 202-343-4747) supplies free information and maps for its national parks, forests, seashores, and historic sites.

**National Parks and Conservation Association** (1015 31st St. NW, Washington, DC 20007; 202-944-8530) focuses on defending, promoting, and improving the national parks system and can provide information about specific sites and programs.

**Rails-to-Trails Conservancy** (1400 16th St. NW, Washington, DC 20036; 202-797-5400) is dedicated to converting abandoned railroad corridors into public trails. There are more than 130 trails around the country for bikers, hikers, walkers, joggers, horseback riders, and even cross-country skiers. Routes include Ohio's Little Miami Scenic Trail, the Washington and Old Dominion Railroad Regional Park in Virginia, Iowa's Heritage Trail, and Florida's Tallahassee-Saint Mark's Trail.

## Further Reading

❑ *Great Expectations* (P.O. Box 8000-411, Sumas, WA 95295), a magazine geared to independent travelers going to offbeat places, with a focus on the environment.

❑ *Lonely Planet* (Embarcadero West, 112 Linden St., Oakland, CA 94607; 415-893-8555) offers travel survival kits and shoestring guides covering virtually all of the third world.

❑ *Directory of Alternative Travel Resources* by Dianne Brause (One World Family Network, Lost Valley Center, 81868 Lost Valley Lane, Dexter, OR 97431; 503-937-3351).

❑ *Your Guide to Environmentally-Sound Boating* (Oceanic Alliance, San Francisco Bay Chapter, Fort Mason, Bldg. E, San Francisco, CA 94123; 415-441-5970).

# *Part III*

# H·O·W T·O G·E·T I·N·V·O·L·V·E·D

# HOW TO GET INVOLVED

From letter-writing campaigns to company boycotts to street demonstrations, Americans have always found ways to get involved in the issues of the day. And when it comes to the environment, there is no shortage of issues on which you can get involved and make your voice heard.

The first step before getting involved is often contacting and supporting one or more of the organizations working on the topics about which you are concerned. (See page 287 for names, addresses, and brief descriptions of the key environmental groups. Many of these groups also have local chapters.) Most of these organizations can provide you with additional resources for understanding these issues, and may offer helpful suggestions on ways you can be effective in making change happen. While you usually need not be a member to obtain information, most memberships are relatively inexpensive—about $25—and besides being tax deductible, may entitle you to additional information or to discounts on literature.

Don't be discouraged by the fact that you "are only one person" or that your tiny local chapter of a big national environmental organization may not seem very powerful. Increasingly, much of the power of environmental change has come from concerned individuals and from local community groups taking action on issues close to home. As the number of such actions has grown, so, too, has the message to national and local government and to corporations: Americans are more concerned than ever about the state of the environment, and they intend to do something about it.

## What the Law Says

Weak enforcement notwithstanding, federal laws have made it easier to learn about and take action on sources of local pollution. There are two key laws:

❑ The **Resource Conservation and Recovery Act**, enacted in 1976, required the Environmental Protection Agency to develop a regulatory program to identify and control hazardous wastes. Among other things, the law established a program of federal grants and technical assistance aimed at encouraging states to improve solid waste management. The law also resulted in the regulation of all hazardous materials, from their manufacture to their transportation and, ultimately, their disposal.

But the law has proved to have little effect on either solid or hazardous wastes. The law was supposed to close hundreds of hazardous-waste dumps, for example, but precious few have been closed. And the Environmental Protection Agency has so far failed to develop a plan to clean up any of the hundreds of sites known to be leaking hazardous chemicals into soil, air, and water. But like the "right-to-know" law, this law requires that companies handling hazardous wastes supply certain information to the state or the Environmental Protection Agency, which must make most of this information public. (The exceptions have mostly to do with information that divulges trade secrets.) So, every facility that treats, stores, or disposes of hazardous waste must conduct regular inspections, keep accurate records, report accidents, and provide other information.

For more information about the Resource Conservation and Recovery Act, contact the **Superfund/RCRA Hotline:** 202-382-3000 or 800-424-9346.

❑ The **Emergency Planning and Community Right-to-Know Act** (part of SARA, the Superfund Amendments and Reauthorization Act) was enacted after the 1984 tragedy in Bhopal, India, in which a Union Carbide plant accidentally released methyl isocyanate, killing more than 2,000 people. The resulting law was aimed at avoiding such accidents in this country by informing citizens about potentially dangerous substances in their communities.

The law requires that companies submit a form annually to the Environmental Protection Agency and to state officials divulging how much of each listed chemical it discharged into the air or water or dumped into the ground. Companies required to submit this information include all those with ten or more employees; that fall into one of twenty Standard Industry Classifications; and that manufacture, process, or use an extensive list of chemicals. In addition, the law requires that approximately 3,500 local emergency planning committees be established to plan for and respond to emergencies involving these companies and chemicals.

Information provided by reporting companies has been compiled into a Toxic Release Inventory database, which is available on-line through the National Library of Medicine's computerized information service TOXNET. Accessing this information requires only a computer and a modem. To obtain an application for TOXNET, contact the **National Library of Medicine** (Specialized Information Services Division, 8600 Rockville Pike, Bethesda, MD 20894; 301-496-6193, 800-638-8480). In addition, many public and university libraries subscribe to TOXNET. The Toxic Release Inventory data are also being made available on microfiche and CD-ROM, a computer-readable format.

Citizens and environmental groups have used the disclosures provided by this law to publicize and challenge some of the worst polluters in their communities. In Maine, for example, one group started issuing a "Terrible Ten Campaign," and has set up workshops to teach local people how to use the law. The state of Illinois enacted a law requiring the state to compile the information submitted by Illinois businesses and to publish it in an annual report. Groups in other states are publishing lists of companies that have failed to file the annual chemical inventories as required by the federal law.

Both laws represent milestones in environmental legislation, but both leave a lot to be desired. Not surprisingly, the provisions of each have become bogged down in the regulatory process, with results coming slowly, often years behind schedule. And, in spite of these laws, industrial pollution persists, with some notorious sites spewing pollutants virtually unchecked. In addition to the need to strengthen such laws, it is clear that more action is needed.

## Taking on Local Polluters

One of the most effective ways individuals can get involved is to target local polluters, and to force action that will stop their poisoning of your community. Consider a few cases of recent years:

❑ In the Houston suburb of Lynchburg, Sandra Mayeaux, a homemaker, spearheaded an effort to close two nearby toxic waste incinerators after she counted the number of her neighbors who had died of cancer. Operators of both incinerators—one of them just a half mile down the street from her house—were ultimately charged with a variety of permit violations in the dispersion of cancer-causing chemicals.

❑ In Berlin, New Jersey, citizens forced a local glass factory to adopt measures that reduced the chances of a toxic spill in the event of an accident.

❑ In Louisiana, a group of citizens organized the "Louisiana Great Toxic March" to bring attention to the high cancer rates in an area between New Orleans and Baton Rouge known as "Cancer Alley." After a rally at the state capitol, the marchers followed the Mississippi River for 130 miles to New Orleans, holding rallies, teach-ins, and press conferences at every stop.

❑ In Galveston County, Texas, Rita Carlson, whose group had only ten active members, organized a lawsuit on behalf of more than 1,000 people who required medical treatment after hydrochloric acid leaked from a nearby oil plant.

Unfortunately, most victories do not come quickly or easily. The wheels of justice seem to grind very slowly when it comes to environmental issues. One reason for this is that both local and national environmental laws are fairly weak, and enforcement of them is even weaker. Some companies have continued to pollute openly for years, knowing that it will take years—decades even—for government authorities to make them stop. By then, the small fines that will be levied against the polluters will be a drop in the bucket compared with what it would have cost the company to modernize its facilities or take some other action to stop dumping toxic wastes into the air and water. And when government agencies do respond, they are often outgunned by the high-priced legal staffs of some of these companies, who are adept at finding loopholes in the law or otherwise stalling and confounding the authorities.

More effective action often requires filing a lawsuit. That may sound intimidating to most individuals and local environmental groups, but the time and expense may be worth the trouble. Lawsuits are time-consuming and expensive, to be sure, and are often settled out of court in a compromise that pleases neither party. But even a partial victory over a polluter can have a positive effect on stemming the problem.

One excellent resource, *Making Polluters Pay: A Citizen's Guide to Legal Action and Organizing* by Audrey Owens Moore, describes the nuts and bolts of the legal process in layman's terms. Designed as a step-by-step workbook for community groups and individuals, the book describes the legal, research, study, and medical efforts required to combat a local polluter, including how to gather information, the barriers toxic victims now face, and how to find and work with a lawyer. The workbook ($15 for individuals; $20 for public interest organizations; $40 for libraries, institutions, law firms, and government agencies) is available from **Environmental Action Foundation** (1526 New Hampshire Ave. NW, Washington, D.C. 20036; 202-745-4870).

Another helpful guide is *A Citizen's Toxic Waste Audit Manual* by Ben Gordon and Peter Montague. The guide is designed to help you identify toxic pollutants generated by your community; find out who is importing toxic wastes into your community; pressure local polluters to reduce waste; and fight polluting facilities proposed for your community. The manual describes the process of conducting a waste audit—"an assessment of the types and amounts of chemical wastes that are being emitted by a particular facility to a particular community." The guide is available from **Greenpeace U.S.A. Toxics Campaign**, 1017 W. Jackson Blvd., Chicago, IL 60607; 312-666-3305. The 72-page guide is free, although Greenpeace requests a $5 donation to defray expenses.

## Other Ways to Get Involved

Lawsuits against local polluters are only one type of action that environmentally concerned citizens can take. Here are ten additional ways to get involved in environmental issues.

❑ **Monitor legislation.** All of the key environmental organizations keep tabs on state and national laws being considered in order to offer testimony and generate letter-writing campaigns on behalf of (or against) proposed laws. Three national organizations have established telephone legislative hotlines to keep interested individuals apprised of bills before Congress: the **Audubon Society** (202-547-9017), **Sierra Club** (202-547-5550), and the **National Wildlife Federation** (202-797-6655). Each hotline offers a two- to four-minute recording.

❑ **Write letters.** It may not seem like a potent weapon, but letters to state and federal legislators on pending bills *do* influence their opinions. According to the National Wildlife Federation, the largest conservation organization in the United States, letter writing helped pass both the 1988 Endangered Species Act and the Clean Water Act. By keeping in touch with local and national environmental organizations, you will know which legislators to write to about what subject, thereby making sure your letters are timely and have maximum impact. Write to senators at the U.S. Senate, Washington, DC 20510; write to members of the House at the U.S. House of Representatives, Washington, DC 20515. Write to the president at the White House, Washington, DC 20500. Use your local public library for names and addresses of state and local officials.

When writing to any public official, keep your letter simple. Focus on one subject and identify a particular piece of legislation, if appropriate. Request a specific action ("Please vote in favor of SB 26890") and state your reasons for your position. If you live or work in the legislator's district, make sure to say so. If you write to a legislator from another district, send a copy of the letter to your own legislator. Keep the letter to one or two paragraphs, never more than one page. Unemotional, courteous letters work best. You might send a copy of the letter to a local newspaper.

In addition to writing letters, you might telephone the office of your legislator. All members of Congress have one or more local offices in their home district, although it probably is more effective to telephone their main office in Washington, D.C. (The main switchboard for Congress is 202-224-3121; operators can connect you to any House or Senate office.) Don't expect to talk to the legislator directly, of course. Ask to speak with the person monitoring environmental issues, state your position in under one minute, thank the individual for speaking with you, and say good-bye.

One handy resource is the "Activist Kit" from **National Wildlife Federation** (1400 16th St. NW, Washington, DC 20036; 202-797-6800). The $5.95 booklet provides information helpful in contacting members of Congress and officials in federal agencies on conservation issues.

❑ **Educate others.** You can do this in a variety of ways, from talking to your friends, co-workers, and neighbors to organizing an educational activity. One such activity might be a "stream walk," a group hike during which participants try to diagnose potential problems. Walking a stream with a knowledgable leader can alert you to erosion problems, highway and construction runoff, excessive algal growth, poisoned fish, foul smells, and direct discharges into the stream. Each of these is cause for concern. The stream walk might be followed by a stream cleanup. Be sure to invite reporters from newspapers and radio and television stations to these activities; their reports will help educate others.

❑ **Campaign for environmental candidates.** Don't just be concerned about someone claiming to be an "environmental president." Look at the environmental positions of candidates at all levels of government. In fact, local candidates—mayor, city council, county supervisors—will probably have a more immediate impact on policies and programs that directly affect the environment in your area. On the national level, **Environmental Action** (1526 New Hampshire Ave. NW, Washington, DC 20036; 202-745-4870) has established an environmentally oriented political action committee called EnAct/PAC, which supports pro-environment candidates for Congress and the White House through lobbying and fundraising.

❑ **Conduct a boycott.** Many environmental groups have sponsored environmentally related boycotts against a number of companies. One useful source for obtaining the latest information is *National Boycott Newsletter* (6506 28th Ave. NE, Seattle, WA 98115; 206-523-0421), issued roughly twice a year, offering nearly 200 pages of detailed information about consumer boycotts (and *buy*cotts—campaigns to support products that have "shown themselves to be particularly ethical, responsible, and courageous"). Subscriptions of four issues are $10 for individuals, $15 for organizations, and $20 for corporations.

You may want to initiate a boycott on the local level as well—against a local carry-out or fast-food outlet, for example, which uses foam cups and containers, or against some other business that generates high volumes of trash but does not attempt to recycle any of it. Advises Environmental Action: "A successful boycott identifies specific companies and formulates clear and specific reform demands. It is not enough, however, to reveal corporate shortcomings; a boycott must mobilize widespread public support." You are advised to be careful when disseminating information about a particular company: if the information is incorrect or misleading, you could be sued by the target company.

❑ **Launch a campaign at work or school.** At Rutgers University, for example, members of the law association decided to target the use of plastic foam in the cafeterias. After creating a multistep, long-term strategy, the students first approached the food services department. The director of food services readily committed to getting rid of foam cups in a matter of days, and the foam food containers as soon as the current inventory was depleted. One of the lessons the students learned: Sometimes, all you have to do is ask. Another school-based activity took place at Brown University, where students saved the school about $2,000 by promoting energy conservation and awareness.

The workplace—whether an office or factory—also offers many opportunities for promoting environmental issues. Recycling paper, bottles, cans, cardboard, and other materials is one good place to start. Removing undesirable materials from the company cafeteria or snack bar is another good effort. Getting the entire company to take on a pet environmental project—cleaning up a local stream, for example, or creating a global-warming campaign—can educate both employees and the community at large. (See also "Making Your Business Green.")

❑ **Promote community recycling.** One of the most common ways to form a recycling organization is to organize a meeting of friends, neighbors, concerned individuals, and members of civic groups interested in recycling. You can hand out fliers throughout the neighborhood and post notices in public spaces to invite people to attend your meeting.

To promote existing recycling opportunities in your community, you can organize a Recycling Day to bring people to the local recycling center. You can also make presentations to local schools and civic groups to promote existing recycling opportunities in your community. Advertisements or articles in local newspapers can facilitate more involvement in local recycling activities. (If you are not sure of the location of the nearest recycling center, check the Yellow Pages under "Recycling Centers" or contact a local trash-hauling company—listed in the Yellow Pages under "Rubbish and Garbage Removal.")

To start a neighborhood recycling center, you need to speak with local salvage and recycling companies, representatives of the beverage industry, and local garbage haulers. All of these groups can help facilitate the collection and transportation of recyclables. Through lobbying, public education, and media work, community recycling groups can be effective advocates of city- or county-operated curbside recycling programs. For more information on starting a community recycling organization, send for the Greenpeace Action Community Recycling Start-up Kit, available from **Greenpeace Action**, 1436 U St. NW, Washington, DC 20009; 202-462-1177. Another helpful resource is "Why Waste a Second Chance?" a motivational video and accompanying guidebook produced by the **National Association of Towns and Townships** (1522 K St. NW, Suite 730, Washington, DC 20005; 202-737-5200). The video program is intended to help community leaders and other concerned citizens generate interest and support for recycling.

❑ **Plant trees.** We've already discussed the many benefits of trees on your community and on the environment. (See "How Trees Save the Earth.") Setting up a tree-planting program is relatively easy. There are a number of organizations that can help organize a one-day or longer-term community campaign to educate citizens on a wide range of environmental issues while assisting them in finding suitable sites, then planting and caring for saplings. Contributors of $15 or more to **Global ReLeaf** (c/o American Forestry Association, 1516 P St. NW, Washington, DC 20005; 202-667-3300) receive a free *Global ReLeaf Action Guide,* containing background information, an action checklist, and access to other Global ReLeaf resources.

❑ **Invite speakers to your organization.** Most environmental organizations offer speakers on a wide range of topics who will speak at no charge to your civic, school, religious, or social organization. Many will also provide literature and additional information. For maximum impact, consider scheduling a debate or panel discussion among representatives of environmental groups, government agencies, and industry. You might turn a seemingly simple topic into a lively discussion. Be sure to invite local media to attend any such event, increasing the number of individuals the message will reach.

❑ **Get involved with government.** Most communities offer a variety of boards, commissions, and committees that deal with environmental issues: planning commissions, zoning and land-use commissions, parks commissions, transit boards, and so on. Each can play a role in setting policies that affect the quality of the environment in your area. For information on how to participate in any of these organizations, contact such organizations directly. Most such groups encourage participation by as many citizens as possible. For information on specific agencies and organizations in your area, contact the local branch of the League of Women Voters or any local or national environmental organization.

## Investing in the Environment

Still another way to be a Green Consumer is to place your finances and investments in environmentally oriented institutions. You need not be a high roller to do this. Many of the organizations listed below have plans for those with as little as $100. All of these organizations have established policies that invest funds in environmentally and socially responsible companies. You are encouraged to contact any of these groups to obtain their latest brochures, applications, and offering statements.

The credit card companies listed have policies that donate a small portion of each dollar charged to the sponsoring organization, offering credit card users the chance to contribute to environmental groups on an ongoing basis.

The final list includes organizations and publications that provide information or assistance to investors on corporate social responsibility and in making socially responsible investments.

Keep in mind that you should always consult a financial adviser before making any investment. Another helpful resource is *A Socially Responsible Financing Planning Guide*, available free to members of **Co-Op America** (2100 M St. NW, Suite 310, Washington, DC 20036; 202-872-5307). See page 297 for more information on Co-Op America.

### Money Market and Mutual Funds

Calvert Social Investment Fund
4550 Montgomery Ave.
Bethesda, MD 20814
800-368-2750

Freedom Environmental Fund
One Beacon St.
Boston, MA 02108
800-225-6258
800-392-6037 in Massachusetts

New Alternatives Fund
295 Northern Blvd.
Great Neck, NY 11021
516-466-0808

Parnassus Fund
244 California St., Suite 210
San Francisco, CA 94111
415-362-3505

Pax World Fund
224 State St.
Portsmouth, NH 03801
603-431-8022

Self-Help Credit Union
413 Chapel Hill St.
Durham, NC 27701
919-683-3016

South Shore Bank
71st and Jeffrey Blvd.
Chicago, IL 60649
312-288-1000

Working Assets Money Fund
230 California St.
San Francisco, CA 94111
415-989-3200, 800-533-3863

## Individual Retirement Accounts (IRAs)

Calvert Social Investment Fund
4550 Montgomery Ave., Suite 1000N
Bethesda, MD 20814
800-368-2750

Pax World Fund
224 State St.
Portsmouth, NH 03801
603-431-8022

Self-Help Credit Union
413 Chapel Hill St.
Durham, NC 27701
919-683-3016

South Shore Bank
71st and Jeffrey Blvd.
Chicago, IL 60649
312-288-1000

Working Assets Money Fund
230 California St.
San Francisco, CA 94111
415-989-3200, 800-533-3863

## Credit Cards

American Rivers
801 Pennsylvania Ave. SE
Washington, DC 20003
202-547-6900 (main office)
704-488-2175 (credit card information)

Co-Op America
2100 M St. NW, Suite 310
Washington, DC 20036
800-424-2667

Defenders of Wildlife
1244 19th St. NW
Washington, DC 20036
202-659-9510, 800-972-9979

International Fund for Animal Welfare
P.O. Box 193
411 Main St.
Yarmouth Port, MA 02675
508-362-4944, 800-972-9979

National Business Association Credit Union
3807 Otter St., P.O. Box 2206
Bristol, PA 19007
800-441-0878

National Wildlife Federation
1400 16th St. NW
Washington, DC 20036
202-797-6800, 800-847-7378

Working Assets
230 California St.
San Francisco, CA 94111
415-989-3200, 800-533-3863

## Organizations and Publications

Catalyst
64 Main St., 2nd Floor
Montpelier, VT 05602
802-223-7943

The Catalyst Group
139 Main St., Suite 614
Brattleboro, VT 05301
802-254-8144

Clean Yield Publications
Box 1880
Greensboro Bend, VT 05842
802-533-7178

Council on Economic Priorities
30 Irving Pl.
New York, NY 10003
212-420-1133

Directory of Environmental Investing
Environmental Economics
1026 Irving St.
Philadelphia, PA 19017
215-925-7168

Directory of Socially Responsible Investments
Funding Exchange
666 Broadway, 5th Floor
New York, NY 10012
212-529-5300

Insight
711 Atlantic Ave., 5th Floor
Boston, MA 02111
617-423-6655

Good Money Publications, Inc.
P.O. Box 363
Worcester, VT 05682
802-223-3911, 800-535-3551

Interfaith Center on Corporate Responsibility
475 Riverside Dr., Room 566
New York, NY 10115
212-870-2936

Investor Responsibility Research Center
1755 Massachusetts Ave. NW
Washington, DC 20036
202-939-6500

## Making Your Business Green

The idea that what's good for the environment is good for business is not exactly new, but it is one that has gained considerable adherents in recent years. Previously limited to smaller companies like **Ben & Jerry's, Tom's of Maine,** and **The Body Shop,** the list of environmentally responsible companies includes such Fortune 500 companies as **Clorox, S.C. Johnson Co.,** and **H.P. Fuller Co.**

In 1989, a coalition of environmental- and social-responsibility groups organized the **Coalition for Environmentally Responsible Economies** (or CERES, after the Roman goddess of agriculture). CERES's first act was to publish the Valdez Principles, a set of standards "for evaluating activities by corporations that directly or indirectly impact the earth's biosphere." The idea of the principles is "to create a voluntary mechanism of corporate self-governance that will maintain business practices consistent with the goals of sustaining our fragile environment for future generations, within a culture that respects all life and honors its interdependence," according to a CERES document.

Here is a summary of the Valdez Principles:

❑ **Protection of the biosphere:** Companies will minimize the release of any pollutant that may damage the air, water, or earth, including those that contribute to the greenhouse effect, depletion of the ozone layer, acid rain, and smog.
❑ **Sustainable use of natural resources:** Companies will make sustainable use of renewable natural resources, such as water, soils, and forests, including protection of wildlife habitat, open spaces, and wilderness, and preservation of biodiversity.
❑ **Reduction and disposal of waste:** Companies will minimize waste, especially hazardous waste, and recycle whenever possible. All waste will be disposed of safely and responsibly.
❑ **Wise use of energy:** Companies will make every effort to use environmentally safe and sustainable energy sources and invest in energy efficiency and conservation.
❑ **Risk reduction:** Companies will minimize environmental and health and safety risks to employees and local communities by employing safe technologies and by being prepared for emergencies.

❑ **Marketing of safe products and services:** Companies will sell products or services that minimize adverse environmental impact and that are safe for consumer use.
❑ **Damage compensation:** Companies will take responsibility for any harm caused to the environment through cleanup and compensation.
❑ **Disclosure:** Companies will disclose to employees and community incidents that cause environmental harm or pose health or safety hazards.
❑ **Environmental directors and managers:** At least one member of the board of directors will be qualified to represent environmental interests, including funding senior executive position for environmental affairs.
❑ **Assessment and annual audit:** Companies will conduct annual self-evaluation of progress in implementing these principles and make results of independent environmental audits available to the public.

For more information about the Valdez Principles, contact the **Social Investment Forum**, 711 Atlantic Ave., Boston, MA 02111; 617-451-3252.

## The Green Bookshelf

### General Environment

*Ages of Gaia: Biography of Our Living Earth,* by James Lovelock (W.W. Norton & Co., 500 5th Ave., New York, NY 10110; 212-354-5500, 800-223-2584. 1988; $16.95).

*Beyond Spaceship Earth: Environmental Ethics and the Solar System,* ed. by Eugene C. Hargrove (Sierra Club Books/Random House, 201 E. 50th St., New York, NY 10022; 212-751-2600, 800-726-0600. 1986; $25).

*Blueprint for a Green Planet,* by John Seymour and Herbert Girardet (Prentice Hall, Prentice Hall Bldg., Sylvan Ave., Englewood Cliffs, NJ 07632; 201-592-2000. 1987; $14.95).

*Blueprint for the Environment: A Plan for Action,* ed. by T. Allan Comp (Howe Brothers Publishers, P.O. Box 6394, Salt Lake City, UT 84106; 801-485-7409, 800-426-5387. 1989; $13.95).

*The Brave Cowboy,* by Edward Abbey (University of New Mexico Press, Journalism Bldg., Albuquerque, NM 87131; 505-277-2346. 1977; $10.95).

*The Crucial Decade: The 1990s and the Global Environmental Challenge* (World Resources Institute, 1750 New York Ave. NW, Washington, DC 20006; 202-393-4055. 1989; $5).

*Cry of the Kalahari,* by Mark and Delia Owens (Houghton Mifflin Co., 1 Beacon St., Boston, MA 02108; 617-725-5000, 800-225-3362. 1985; $7.95).

*Deep Ecology,* by Bill Devall and George Sessions (Gibbs Smith, Publisher/Peregrine Smith Books, P.O. Box 667, Layton, UT 84041; 801-544-9800, 800-421-8714. 1987; $11.95).

*Diet for a Small Planet,* by Frances Moore Lappe (Ballantine, 201 E. 50th St., New York, NY 10022; 212-751-2600, 800-638-6460. 1985; $4.95).

*Earth and Other Ethics*, by Christopher D. Stone (Prentice Hall, Prentice Hall Bldg., Sylvan Ave., Englewood Cliffs, NJ 07632; 201-592-2000. 1987; $7.95).

*The Earth Care Annual* (Rodale Press/St. Martin's Press, 175 Fifth Ave., New York, NY 10010; 212-674-5151, 800-221-7945. 1990; $17.95).

*The Earth Report: The Essential Guide to Global Ecological Issues*, ed. by Edward Goldsmith and Nicholas Hildyard (Price Stern Sloan, 360 N. La Cienega Blvd., Los Angeles, CA 90048; 213-657-6100, 800-421-0892. 1988; $12.95).

*Ecotopia*, by Ernest Callenbach (Bantam, 666 Fifth Ave., New York, NY 10103; 212-765-6500, 800-223-6834. 1983; $4.50).

*Elephant Memories: Thirteen Years in the Life of an Elephant Family*, by Cynthia Moss (Fawcett, 201 E. 50th St., New York, NY 10022; 212-751-2600, 800-726-0600. 1989; $10.95).

*The Endangered Kingdom: The Struggle to Save America's Wildlife*, by Riger L. DiSilvestro (John Wiley & Sons, 605 Third Ave., New York, NY 10158; 212-850-6222. 1989; $19.95).

*Environmental Ethics*, by Holmes Tolston (Temple University Press, 1601 N. Broad St., USB 306, Philadelphia, PA 1912; 215-787-8787. 1988; $16.95 paper, $39.95 cloth).

*50 Simple Things You Can Do to Save the Earth*, by The Earth Works Group (Earthworks Press, 1400 Shattuck Ave., Box 25, Berkeley, CA 94709. 1989; $4.95).

*The Fragile Environment*, ed. by Laurie Friday and Ronald Laskey (Cambridge University Press, 32 E. 57th St., New York, NY 10022; 212-688-8885, 800-221-4512. 1989; $19.95).

*Gaia: An Atlas of Planet Management*, by Norman Myers (Anchor Press/Doubleday, 666 Fifth Ave., New York, NY 10103; 212-765-6500, 800-223-6834. 1984; $18.95).

*How to Make the World a Better Place: A Beginner's Guide*, by Jeffrey Hollender (William Morrow & Co., 105 Madison Ave., New York, NY 10016; 212-889-3050, 800-843-9389. 1990; $9.95 paper).

*Jacques Cousteau's Amazon Journey*, by Jacques-Yves Cousteau and Mose Richards (Harry N. Abrams Publishers, 100 Fifth Ave., New York, NY 10011; 212-206-7715, 800-345-1359. 1984; $39.95).

*John McPhee Reader*, ed. by William L. Howarth (Farrar, Straus & Giroux, 19 Union Sq. W., New York, NY 10003; 212-741-6900, 800-242-7737. 1976; $9.95).

*The Last Extinction*, ed. by Les Kaufman and Kenneth Malloy (MIT Press, 55 Hayward St., Cambridge, MA 02138; 617-253-5646. 1986; $16.95).

*Life in the Balance*, by David Rains Wallace (National Audubon Society, 950 Third Ave., New York, NY 10022; 212-832-3200. 1987; $29.95).

*The Long Shadowed Forest*, by Helene Hoover (W.W. Norton & Co., 500 Fifth Ave., New York, NY 10110; 212-354-5500, 800-223-2584. 1980; $4.95).

*Nature's End: The Consequences of the Twentieth Century*, by Whitley Streiber and James Kunetka (Warner Books, 666 Fifth Ave., New York, NY 10103; 212-484-2900. 1987; $4.95).

*The New Environmental Age*, by Max Nicholson (Cambridge University Press, 32 E. 57th St., New York, NY 10022; 212-688-8885, 800-221-4512. 1987; $27.95).

*New World, New Mind*, by Paul Ehrlich and Robert Ornstein (Doubleday, 666 Fifth Ave., New York, NY 10103; 212-765-6500, 800-223-6834. 1989; $18.95).

*News of the Universe*, ed. by Robert Bly (Sierra Club Books/Random House, 201 E. 50th St., New York, NY 10022; 212-751-2600; 800-726-0600. 1980; $8.99).

*The Peregrine,* by John Baker (University of Idaho Press, Moscow, ID 83843; 208-885-6245. 1986; $10.95).

*The Politics of the Solar Age: Alternatives to Economics,* by Hazel Henderson (Anchor Press/Doubleday, 666 Fifth Ave., New York, NY 10103; 212-765-6500, 800-223-6834. 1981; $15.95).

*Restoring the Earth: How Americans Are Working to Renew Our Damaged Environment,* by John J. Berger (Alfred A. Knopf, 201 E. 50th St., New York, NY 10022; 212-751-2600; 800-726-0600. 1985; $18.95).

*The Rights of Nature: A History of Environmental Ethics,* by Roderick Frazier Nash (University of Wisconsin Press, 114 N. Murray St., Madison, WI 53715; 608-262-4928. 1989; $27.50).

*The Rivers Amazon,* by Alex Shoumatoff (Sierra Club Books/Random House, 201 E. 50th St., New York, NY 10022; 212-751-2600, 800-726-0600. 1986; $8.95).

*Simple in Means, Rich in Ends: Practicing Deep Ecology,* by Bill Devall (Gibbs Smith, Publisher, P.O. Box 667, Layton, UT 84041; 801-544-9800, 800-421-8714. 1988; $12.95)

*State of the Ark: An Atlas of Conservation in Action,* by Lee Durrell (Doubleday, 666 Fifth Ave., New York, NY 10103; 212-765-6500, 800-223-6834. 1986; $14.95).

*State of the Environment: A View Toward the Nineties,* by the Conservation Foundation (Conservation Foundation/Island Press, 1718 Connecticut Ave. NW, Suite 300, Washington, DC 20009; 202-232-7933. 1987; $19.95).

*State of the States* (Renew America, 1001 Connecticut Ave. NW, Suite 719, Washington, DC 20036; 202-232-2252. annual; $20).

*State of the World* (Worldwatch Institute, 1776 Massachusetts Ave. NW, Washington, DC 20036; 202-452-1999. annual; $9.95).

*Toxics, Chemicals, Health, and the Environment*, ed. by Lester B. Lave and Arthur C. Upton (The Johns Hopkins University Press, Baltimore, MD 21211; 301-338-6900. 1987; $16.50).

*The Toxic Cloud*, by Michael H. Brown (Harper & Row, 10 E. 53rd St., New York, NY 10022; 212-207-7000, 800-242-7737. 1987; $18.95).

*Whatever Happened to Ecology?*, by Stephanie Mills (Sierra Club Books/Random House, 201 E. 50th St., New York, NY 10022; 212-751-2600, 800-726-0600. 1989; $18.95).

*Who's Poisoning America: Corporate Polluters and Their Victims in the Chemical Age*, ed. by Ralph Nader, Ronald Brownstein, and John Richard (Sierra Club Books/Random House, 201 E. 50th St., New York, NY 10022; 212-751-2600, 800-726-0600. 1982; $12.95).

*World Resources* (World Resources Institute/International Institute for Environment and Development/United Nations Environment Programme, 1750 New York Ave. NW, Washington, DC 20006; 202-393-4055. annual; $16.95).

## Air Pollution

*A Killing Rain: The Global Threat of Acid Precipitation*, by Thomas Pawlick (Sierra Club Books/Random House, 201 E. 50th St., New York, NY 10022; 212-751-2600, 800-726-0600. 1988; $15.95).

*Air Pollution's Toll on Forests and Crops*, ed. by James J. MacKenzie and Mohamed T. El-Ashry (World Resources Institute, 1750 New York Ave. NW, Washington, DC 20006; 202-638-6300. 1989; $46).

*Breathing Easier: Taking Action on Climate Change, Air Pollution, and Energy Insecurity*, by James J. MacKenzie (World Resources Institute, 1750 New York Ave. NW, Washington, DC 20006; 202-638-6300. 1988; $5).

*Climate Change and Acid Rain*, by Sheila Machado and Rick Piltz (Renew America, 1001 Connecticut Ave. NW, Suite 719, Washington, DC 20036; 202-232-2252. 1988; $10).

*Who's Who of American Toxic Air Polluters: Guide to More than 1500 Factories in 46 States Emitting Cancer-Causing Chemicals,* by the Natural Resources Defense Council (NRDC, 1350 New York Ave. NW, Washington, DC 20005; 202-783-7800. 1989; $25).

## Animal Cruelty/Animal Rights

*Animal Liberation,* by Peter Singer (Prentice-Hall, Prentice-Hall Bldg., Sylvan Ave., Englewood Cliffs, NJ 07632; 201-592-2000. 1989; $14.95).

*Animals, Nature and Albert Schweitzer,* ed. by Ann Cottrell Free (Humane Society of the United States, 2100 L St. NW, Washington, DC 20037; 202-452-1100. 1982; $3.50).

## Automobiles

*Corporate Crime and Violence,* by Russell Mokhiber (Sierra Club Books/Random House, 201 E. 50th St., New York, NY 10022; 212-751-2600, 800-726-0600. 1988; $25).

*Rethinking the Role of the Automobile,* by Michael Renner (Worldwatch Institute, 1776 Massachusetts Ave. NW, Washington, DC 20036; 202-452-1999. 1988; $4).

## Composting

*The Art of Composting* (Metropolitan Service District, 2000 S.W. 1st Ave., Portland, OR 97201; 503-221-1646. 1989; free).

*Let It Rot! The Home Gardener's Guide to Composting,* by Stu Campbell (Garden Way Publishing/Storey Communications Inc., Schoolhouse Rd., Pownal, VT 05261; 802-823-5811, 800-441-5700. 1975; $5.95).

*The Rodale Guide to Composting,* by Jerry Minnich, Marjorie Hunt, et al. (Rodale Press, 33 E. Minor St., Emmaus, PA 18049; 215-967-5171, 800-247-5028. 1979; $14.95).)

## Energy

*Beyond Oil: The Threat to Food and Fuel in the Coming Decades*, by John Gerver et al. (Ballinger, 54 Church St., Cambridge, MA 02142; 617-492-0670, 800-242-7737. 1986; $17.95).

*Energy: A Guidebook*, by Janet Ramage (Oxford University Press, 200 Madison Ave., New York, NY 10011; 212-679-7300, 800-451-7556. 1983; $16.50).

*Energy Conservation: A Campus Guidebook*, by Kevin O'Brien and David Corn (Center for Study of Responsive Law, P.O. Box 19367, Washington, DC 20036; 202-387-8030. 1981; $5).

*Energy for a Sustainable World*, by Jose Goldemberg et al. (World Resources Institute, 1750 New York Ave. NW, Washington, DC 20006; 202-638-6300. 1987; $10).

*Energy for Survival*, ed. by Harry Messel (Pergamon Press, Maxwell House, Fairview Park, Elmsford, NY 10523; 914-592-7700. 1979; $16.50).

*Energy Future*, ed. by Robert Stobaugh and Daniel Yergin (Random House, 201 E. 50th St., New York, NY 10022; 212-751-2600, 800-726-0600. 1982; $7.95).

*Energy Unbound: A Fable for America's Future*, by L. Hunter Lovins, Amory B. Lovins, and Seth Zuckerman (Sierra Club Books/Random House, 201 E. 50th St., New York, NY 10022; 212-751-2600, 800-726-0600. 1985; $17.95).

*The Home Energy Decision Book*, by Gigi Coe, Michael Garland, and Michael Eaton (Sierra Club Books/Random House, 201 E. 50th St., New York, NY 10022; 212-751-2600, 800-726-0600. 1984; $9.95).

*More Other Homes and Garbage: Designs for Self-Sufficient Living*, by Jim Leckie, Gil Masters, Harry Whitehouse, and Lily Young (Sierra Club Books/Random House, 201 E. 50th St., New York, NY 10022; 212-751-2600, 800-726-0600. 1982; $14.95).

*Rays of Hope: The Transition to a Post-Petroleum World*, by Denis Hayes (W.W. Norton & Co., 500 Fifth Ave., New York, NY 10110; 212-354-5500, 800-223-2584. 1977; $4.95).

*Resource-Efficient Housing Guide*, by Robert Sardinsky (Rocky Mountain Institute, 1739 Snowmass Creek Rd., Snowmass, CO 81654; 303-927-3851. 1989; $15 postpaid).

*Your Affordable Solar Home*, by Dan Hibshman (Sierra Club Books/Random House, 201 E. 50th St., New York, NY 10022; 212-751-2600, 800-726-0600. 1983; $7.95).

## Food Safety

*For Our Kids' Sake: How to Protect Your Child Against Pesticides in Food*, by the Natural Resources Defense Council (Sierra Club Books/Random House, 201 E. 50th St., New York, NY 10022; 212-751-2600, 800-726-0600. 1989; $6.95).

*A Marketbasket of Food Hazards: Critical Gaps in Food Protection*, (Public Voice for Food & Health Policy, 1001 Connecticut Ave. NW, Suite 522, Washington, DC 10036; 202-659-5930. 1983; $12).

## Garbage

*Complete Trash: The Best Way to Get Rid of Practically Everything Around the House*, by Norm Crampton (Little, Brown & Co., 200 West St., Waltham, MA 02154; 617-890-0250, 800-343-9204. 1989; $8.70).

*Rush to Burn: Solving America's Garbage Crisis?* from Newsday (Island Press, 1718 Connecticut Ave. NW, Suite 300, Washington, DC 20009; 202-232-7933. 1989; $14.95 paper, $22.95 cloth).

*War on Waste: Can America Win Its Battle with Garbage?*, by Louis Blumberg and Robert Gottlieb (Island Press, 1718 Connecticut Ave. NW, Suite 300, Washington, DC 20036; 202-232-7933. 1989; $19.95 paper, $34.95 cloth).

## Gardening and Landscaping

*The Basic Book of Organic Gardening,* ed. by Robert Rodale (Ballantine, 201 E. 50th St., New York, NY 10022; 212-751-2600, 800-726-0600. 1987; $3.95).

*Bio-Dynamic Farm: Agriculture in the Service of the Earth and Humanity,* by Herbert H. Koepf (Anthroposophic Press Inc., Bells Pond, Star Rt., Hudson, NY 12534; 518-851-2054. 1989; $11.95).

*Community Open Spaces: Green Neighborhoods through Community Action and Land Conservation,* by Mark Francis, Lisa Cashdan, and Lynn Paxson (Island Press, 1718 Connecticut Ave. NW, Suite 300, Washington, DC 20036; 202-232-7933. 1984; $24.95).

*Companion Plants and How to Use Them,* by Helen Philbrick and Richard B. Gregg (Devin-Adair Publishers, 6 N. Water St., Greenwich, CT 06830; 203-531-7755. 1966; $6.95).

*The Complete Book of Edible Landscaping: Home Landscaping with Food-Bearing Plants and Resource-Saving Techniques,* by Rosalind Creasy (Sierra Club Books/Random House, 201 E. 50th St., New York, NY 10022; 212-751-2600, 800-726-0600. 1982; $19.95 paper).

*The Encyclopedia of Natural Insect and Disease Control,* ed. by Roger B. Yepsen, Jr. (Rodale Press, 33 E. Minor St., Emmaus, PA 18049; 215-967-5171, 800-247-5028. 1984; $24.95).

*How to Grow Vegetables Organically,* by Jeff Cox et al (Rodale Press, 33 E. Minor St., Emmaus, PA 18049; 215-967-5171, 800-441-7761. 1988; $21.95)

*The Natural Garden,* by Ken Druse (Crown Publishers, 201 E. 50th St., New York, NY 10022; 212-751-2600, 800-726-0600. 1987; $35).

*The Organic Garden Book,* by Geoff Hamilton (Crown Publishers, 201 E. 50th St., New York, NY 10022; 212-751-2600, 800-726-0600. 1987; $27.50).

*Saving Water in a Desert City*, by William E. Martin, Helen M. Ingram, Nancy K. Laney, and Adrian H. Griffin (Resources for the Future, 1616 P St. NW, Washington, DC 2003;, 202-328-5086. 1984; $10).

## Getting Involved

*Adopting a Stream: A Northwest Handbook*, by Steve Yates (University of Washington Press, P.O. Box 50096, Seattle, WA 98145; 209-543-8870, 800-441-4115. 1988; $9.95).

*Beyond 25 Percent: Materials Recovery Comes of Age*, by Theresa Allen, Brenda Platt, and David Morris. (Institute for Local Self-Reliance, 2425 18th St. NW, Washington, DC 20009; 202-232-4108. 1987; $40; $25 for grassroots groups).

*Citizen Suits: Private Enforcement of Federal Pollution Control Laws*, by Jeffrey G. Miller (John Wiley & Sons/Island Press, 1718 Connecticut Ave. NW, Suite 300, Washington, DC 20036; 202-232-7933. 1987; $85).

*The Complete Guide to Environmental Careers*, by the CEIP Fund (Island Press, 1718 Connecticut Ave. NW, Suite 300, Washington, DC 20036; 202-232-7933. 1989; $14.95 paper, $24.95 cloth).

*Love Canal: My Story*, by Lois Marie Gibbs (State University of New York Press, State University Plaza, Albany, NY 12246; 518-472-5000, 800-252-3206. 1982; $14.95).

*Making Polluters Pay: A Citizens' Guide to Legal Action and Organizing*, by Andrew Owens Moore (Environmental Action, 1525 New Hampshire Ave. NW, Washington, DC 20036; 202-745-4870. 1987; $15 for individuals, $20 for public interest organizations, $40 libraries).

*Raising Hell: A Citizen's Guide to the Fine Art of Investigation*, by Thomas Noyes (Foundation for National Progress, 1663 Mission St., San Francisco, CA 94103; 415-558-8881).

*Superfund Deskbook*, by the Environmental Law Reporter Staff (Environmental Law Institute, 1616 P St. NW, Suite 200, Washington, DC 20036; 202-328-5150. 1986; $75).

## Global Warming

*The Challenge of Global Warming*, ed. by Dean Edwin Abrahamson (Island Press, 1718 Connecticut Ave. NW, Suite 300, Washington, DC 20036; 202-232-7933. 1989; $19.95 paper, $34.95 cloth).

*Climate Change and Society: Consequences of Increasing Atmosphere Carbon Dioxide*, by William Kellogg and Robert Schware (Westview Press, 5500 Central Ave., Boulder, CO 80301; 303-444-3541. 1981; $12).

*Climatic Change*, ed. by John Gribbon (Cambridge University Press, 110 Midland Ave., Port Chester, NY 10573; 914-937-9600, 800-227-0247. 1978; $27.95).

*The End of Nature*, by William McKibben (Random House, 201 E. 50th St., New York, NY 10022; 212-751-2600, 800-726-0600. 1989; $19.95).

*Entropy: Into the Greenhouse World*, by Jeremy Rifkin with Ted Howard (Bantam Books, 666 Fifth Ave., New York, NY 10103; 212-765-6500, 800-223-6834. 1989; $9.95).

*Global Warming*, by Stephen H. Schneider (Sierra Club Books/ Random House, 201 E. 50th St., New York, NY 10022; 212-751-2600, 800-726-0600. 1989; $18.95).

*The Greenhouse Effect and Sea Level Rise*, by Michael C. Barth and James G. Titus (Van Nostrand Reinhold Co., 115 Fifth Ave., New York, NY 10003; 212-254-3232. 1984; $34.95).

*The Greenhouse Gases* (United Nations Environment Program, 1889 F St. NW, Washington, DC 20006; 202-289-8456. 1987; free).

*The Greenhouse Trap*, ed. by Francesca Lyman with Irving Mintzer, Kathleen Courrier, and Jane MacKenzie (Beacon Press, 25 Beacon St., Boston, MA 02108; 617-742-2110. 1990; $9.95).

*The Hole in the Sky*, by John Gribbon (Bantam, 666 Fifth Ave., New York, NY 10103; 212-765-6500, 800-223-6834. 1988; $4.50).

*Ozone Crisis: The 15-Year Evolution of a Sudden Global Emergency*, by Sharon Roan (John Wiley & Sons, 605 Third Ave., New York, NY 10158; 212-850-6000. 1989; $18.95).

*Protecting the Ozone Layer*, by the Environmental Defense Fund (EDF, 257 Park Ave. S., New York, NY 10010; 212-505-2100. 1988; $2).

*Reducing the Rate of Global Warming: The States' Role*, by Sheila Machado and Rick Piltz (Renew America, 1001 Connecticut Ave. NW, Suite 719, Washington, DC 20036; 202-232-2252. 1988; $10).

## Hazardous Waste

*America's Future in Toxic Waste Management*, by Bruce W. Piasecki and Gary A. Davis (Quorum Books/Greenwood Press, 88 Post Rd. W., Westport, CT 06881; 203-226-3571, 800-225-5800. 1987; $49.95).

*Contaminated Communities: The Social and Psychological Impacts of Residential Toxic Exposure*, by Michael R. Edelstein (Island Press, 1718 Connecticut Ave. NW, Suite 300, Washington, DC 20036; 202-232-7933. 1988; $29.95).

*Hazardous Waste: Confronting the Challenge*, by Christopher Harris, William L. Want, and Morris A. Ward (Greenwood Press/Environmental Law Institute, P.O. Box 5007, Westport, CT 06881; 203-226-3571. 1987; $39.95).

*Hazardous Waste in America*, by Samuel S. Epstein (Sierra Club Books/Random House, 201 E. 50th St., New York, NY 10022, 212-751-2600, 800-726-0600. 1983; $12.95).

## Pesticides

*Agricide: The Hidden Crisis That Affects Us All*, by Dr. Michael W. Fox (Schocken Books/Random House, 201 E. 50th St., New York, NY 10022, 212-751-2600, 800-726-0600. 1986; $7.95).

*The Bhopal Syndrome: Pesticides, Environment, and Health,* by David Weir (Sierra Club Books/Random House, 201 E. 50th St., New York, NY 10022, 212-751-2600, 800-726-0600. 1986; $9.95 paper, $17.95 cloth).

*Pesticide Alert: A Guide to Pesticides in Fruits and Vegetables,* by Laurie Mott and Karen Snyder (Natural Resources Defense Council/Sierra Club Books, 201 E. 50th St., New York, NY 10022; 212-751-2600, 800-638-6460. 1988; $6.95 paper, $15.95 cloth).

*Silent Spring,* 25th Anniversary Edition, by Rachel Carson (Houghton Mifflin Co., 1 Beacon St., Boston, MA 02108; 617-725-5000, 800-225-3362. 1987; $7.95).

## Pet Care

*Dr. Pitcairn's Complete Guide to Natural Health for Dogs and Cats,* by Richard H. Pitcairn, D. VM., Ph.D., and Susan Hubble Pitcairn (Rodale Press, 33 E. Minor St., Emmaus, PA 18049; 215-967-5171, 800-247-5028. 1982; $12.95).

## Rain Forests

*Animal Extinctions: What Everyone Should Know,* ed. by R.J. Hoage (Smithsonian Institute Press, 955 L'Enfant Plaza, Room 2100, Washington, DC 20560; 202-287-3738, 800-678-2675. 1985; $11.95).

*Biodiversity,* ed. by E.O. Wilson (Island Press, 1718 Connecticut Ave. NW, Suite 300, Washington, DC 20036; 202-232-7933. 1988; $19.50).

*Forest Primeval: The Natural History of an Ancient Forest,* by Chris Maser (Sierra Club Books/Random House, 201 E. 50th St., New York, NY 10022; 212-751-2600. 800-726-0600. 1989; $25).

*Hoofprints on the Forest: Cattle Ranching and the Destruction of Latin America's Tropical Forests,* by Douglas R. Shane (Institute for the Study of Human Issues, 210 S. 13th St., Philadelphia, PA 19107; 215-732-9729. 1986; $27.50).

*In the Rainforest,* by Catherine Caufield (University of Chicago Press, 5801 Ellis Ave., Chicago, IL 60637; 312-702-7700, 800-621-2736. 1986; $11.95).

*Life Above the Jungle Floor,* Donald Perry (Fireside/Simon & Schuster, Simon & Schuster Bldg., 1230 Avenue of the Americas, New York, NY 10020; 212-698-7000, 800-223-2348. 1988; $11.95).

*People of the Tropical Rain Forest,* ed. by Julie Sloan Denslow and Christine Padoch (University of California Press/Island Press, 1718 Connecticut Ave. NW, Suite 300, Washington, DC 20036; 202-232-7933. 1988; $19.95 paper, $39.50 cloth).

*Saving Tropical Forests,* by Judith Gradwohl and Russell Greenberg (Island Press, 1718 Connecticut Ave. NW, Suite 300, Washington, DC 20036; 202-232-7933. 1988; $24.95).

*Tropical Nature: Life and Death in the Rain Forests of Central and South America,* by Adrian Forsyth and Ken Miyata (Scribner's/Macmillan Publishing Co., 866 Third Ave., New York, NY 10022; 212-702-2000, 800-257-5755. 1987; $7.95).

*The Tropical Rain Forest: A First Encounter,* by Marius Jacobs and R.A.A. Oldeman (Springer-Verlag/Island Press, 1718 Connecticut Ave. NW, Suite 300, Washington, DC 20036; 202-232-7933. 1988, $39.95).

*Tropical Rain Forest Ecology,* by D.J. Mabberly (Routledge Chapman and Hall, 29 W. 35th St., New York, NY 10001; 212-244-3336. 1983; $29.95).

## Recycling

*Coming Full Circle: Successful Recycling Today,* by the Environmental Defense Fund (EDF, 257 Park Ave. S., New York, NY 10010; 212-505-2100. 1988; $20).

*Planning for Community Recycling: A Citizen's Guide to Resources* (Environmental Action, 1525 New Hampshire Ave. NW, Washington, DC 20036; 202-745-4870. 1986; free).

*Promoting Source Reduction and Recyclability in the Marketplace* (EPA, RCRA Docket, 401 M St. SW, Washington, DC 20460; 202-475-9327. 1989; free).

## Socially Responsible Investing

*Economics As If the Earth Really Mattered,* by Susan Meeker-Lowry (New Society Publishers, P.O. Box 582, Santa Cruz, CA 90061; 408-458-1191. 1988; $9.95).

*Ethical Investing,* by Amy Domini (Addison-Wesley Publishers, Route 128, Reading, MA 01867; 617-944-3700, 800-447-2226. 1986; $12.95).

*Rating America's Corporate Conscience,* by the Council on Economic Priorities (CEP, Addison-Wesley Publishing, Route 128, Reading, MA 01867; 617-944-3700, 800-447-2226. 1987; $14.95).

*Socially Responsible Investing,* by Rob Baird (Center for Urban Education, 3835 SW Kelly, Portland, OR 97201; 503-223-3444. 1987; $9.95).

## Water Pollution

*A Citizen's Guide to Plastics in the Ocean,* by Kathryn J. O'Hara and Suzanne Iudicello (Center for Marine Conservation, 1725 DeSales St. NW, Washington, DC 20036, 202-429-5609. 1988; $2 postpaid).

*The Water Planet,* by Lyall Watson (Crown Publishers/Random House, 201 E. 50th St., New York, NY 10022, 212-751-2600, 800-726-0600. 1988; $30).

## Books for Children and Young Adults

*Chadwick the Crab,* by Priscilla Cummings (Tidewater Publishers, Box 456, Centreville, MD 21617, 301-758-1075, 800-638-7641. 1986; $5.95).

*Chadwick and the Garplegrungen,* by Priscilla Cummings (Tidewater Publishers, Box 456, Centreville, MD 21617, 301-758-1075. 1987; $6.95).

*The Lorax,* by Dr. Seuss (Random House, 201 E. 50th St., New York, NY 10022, 212-751-2600, 800-726-0600. 1971; $10.95).

*The Planet of Trash: An Environmental Fable,* by George Poppel (National Press Inc., 7201 Wisconsin Ave., Suite 720, Bethesda, MD 20814; 301-657-1616. 1987; $9.95 postpaid).

*The Wump World,* by Bill Peet (Houghton Mifflin Co., 1 Beacon St., Boston, MA 02108, 617-725-5000. 1974; $3.95).

## Directories of Environmental Organizations

*Conservation Directory,* by the National Wildlife Federation (NWF, 1400 16th St. NW, Washington, DC 20036, 202-797-6800; annual; $15).

*Directory of State Environmental Agencies,* by the Environmental Law Institute (Environmental Law Institute/Island Press, 1718 Connecticut Ave. NW, Suite 300, Washington, DC 20036, 202-232-7933. 1985; $22.50).

## Magazines

In addition to the magazines listed below are many fine periodicals published by environmental organizations; see listings on page 287. An asterisk (*) indicates that magazine is printed on recycled paper.

*Archives of Environmental Health,* Heldreff Publications, 4000 Albemarle St. NW, Washington, DC 20016; 202-362-6445. Annual subscription: for $85 six issues.

*BioCycle: Journal of Waste Recycling,* P.O. Box 351, Emmaus, PA 18049; 215-967-4135. Annual subscription: $44 for twelve issues.

*Buzzworm: The Environmental Journal,* 1818 16th St., Boulder, CO 80302; 303-442-1969. Annual subscription: $18 for six issues.

**E Magazine,* P.O. Box 5098, Westport, CT 06881; 203-854-5559. Annual subscription: $20 for six issues.

*Environment,* Heldreff Publications, 4000 Albemarle St. NW, Washington, DC 20016; 202-362-6445. Annual subscription: $23 for ten issues.

**EPA Journal,* Superintendent of Documents, Government Printing Office, Washington, DC 20402; 202-362-6445. Annual subscription: $8 for six issues.

**Garbage: The Practical Journal for the Environment,* 435 9th St., Brooklyn, NY 11215; 718-788-1700, 800-274-9909. Annual subscription: $21 for six issues.

*Harrowsmith,* Ferry Rd., Charlotte, VT 05445; 802-425-3961. Annual subscription: $24 for six issues.

*Journal of Environmental Education* (Heldreff Publications, 4000 Albemarle St. NW, Washington, DC 20016; 202-362-6445. Annual subscription: $47for four issues.

*One Person's Impact,* P.O. Box 751, Westborough, MA 01581; 508-478-3716. Annual subscription: $24 for six issues and a special report; includes telephone reference service.

## Environmental Organizations

**Acid Rain Foundation** (1410 Varsity Dr., Raleigh, NC 27606; 919-828-9443) is dedicated to developing public awareness, information, educational materials, and research in the area of acid disposition, air pollution and toxins, global change, and recycling. Membership is $25 per year and includes the quarterly publication, *The Acid Rain Update.*

**Acid Rain Information Clearinghouse** (33 S. Washington St., Rochester, NY 14608; 716-546-3796) provides comprehensive reference and referral, current awareness, and educational services to a wide range of professionals, academics, and public interest groups. The organization maintains a library, sponsors conferences and seminars, and prepares topical bibliographies.

**Activist Network** (see **National Audubon Society**)

**Adopt-a-Stream Foundation** (P.O. Box 5558, Everett, WA 98201; 206-388-3313) promotes environmental education and stream enhancement. It offers support and guidance to those "adopting" a stream: The adopting group provides long-term care of the stream. Membership in this nonprofit, volunteer-run organization is $10 and up, and includes a subscription to the quarterly newsletter, *StreamLines*. The foundation publishes two informative books: *Adopting a Stream* and *Adopting a Wetland.*

**African Wildlife Foundation** (1717 Massachusetts Ave. NW, Washington, DC 20036; 202-265-8394) works with Africans in over twenty countries, promoting, establishing, and supporting grassroots and institutional programs in conservation, wildlife management training, and management of threatened conservation areas. The group's current emphasis is on educating Americans not to buy ivory. Membership is by donation; $15 minimum required to subscribe to the quarterly newsletter, *Wildlife News*.

**Alliance for Environmental Education** (2111 Wilson Blvd., Suite 751, Arlington, VA 22201; 703-875-8660) is a nonprofit organization dedicated to promoting the development of environmental education programs. Activities include national conferences, publications, and the "Network for Environmental Education." Membership ($100 per year) is available to organizations and includes a subscription to the bimonthly newsletter *The Network Exchange*.

**Alliance to Save Energy** (1725 K St. NW, Washington, DC 20006; 202-857-0666) is dedicated to increasing energy efficiency. The group conducts research and pilot projects to evaluate solutions to energy-efficiency problems. Alliance programs address quality of life, environment, national security, international competitiveness, and economic development. It is a nonmembership organization supported by corporations, foundations, and other organizations promoting energy efficiency. Publications and computer software are available to the public.

**America the Beautiful Fund** (219 Shoreham Bldg., Washington, DC 20005; 202-638-1649) was organized in 1965 to give recognition, technical support, small seed grants, gifts of free seeds, and national recognition awards to volunteers and community groups that initiate new local projects improving environmental quality. Activities include beautifying communities through seed donations and growing food for the needy. Membership ($5 and up) includes a subscription to *Better Times*, a quarterly newsletter.

**American Cave Conservation Association** (P.O. Box 409, Horse Cave, KY 42749; 502-786-1466) was established to protect and preserve caves, karstlands, and groundwater. The group work focuses on education, creation of a national education facility and museum of caves and karst, and developing an information network. Annual membership fees are $25 and include a bimonthly magazine *American Caves* and a periodic newsletter.

**American Cetacean Society** (P.O. Box 2639, San Pedro, CA 90731; 213-548-6279) is a nonprofit volunteer organization working to protect whales and dolphins through research, conservation, and education. Projects have focused on the killings of dolphins by tuna fishermen, stopping whaling, ocean pollution, strandings, and gill net and drift net problems. Membership is $25 per year and includes a subscription to the quarterly *Whale Watcher* magazine and *Whale News* newsletter.

**American Council for an Energy-Efficient Economy** (1001 Connecticut Ave. NW, Suite 535, Washington, DC 20036; 202-429-8873) gathers, evaluates, and disseminates information to stimulate greater energy efficiency. Topics of focus include buildings, appliances, and indoor air quality. It annually publishes the booklet *The Most Energy Efficient Appliances*. Although not a membership organization, it maintains a mailing list.

**American Forest Council** (1250 Connecticut Ave. NW, Suite 320, Washington, DC 20036; 202-463-2455) is sponsored by the forest products industry to educate and train private forest land owners in good forest-management practices. It sponsors an education program, "Project Learning Tree," through its American Tree Farm System. The council publishes the monthly *American Forest Council* magazine; a poster/magazine *Green America* is available through Project Learning Tree. Membership is available to companies, private forest tree farmers, and individuals.

**American Forestry Association** (P.O. Box 2000, Washington, DC 20013; 202-667-3300) is a national citizens organization dedicated to the maintenance and improvement of the health and value of trees and forests, and to make Americans more aware of and active in forest conservation and tree planting. It sponsors "Global ReLeaf," a national program encouraging Americans to plant millions of trees to lower carbon dioxide levels and beautify communities. Another major program is the "National Register of Big Trees." Publications include the *Resource Hotline* newsletter. Membership is $24 annually and includes a subscription to the bimonthly *American Forests*.

**American Humane Association** (9725 E. Hampden Ave., Denver, CO 80231; 303-695-0811) was founded to protect animals from neglect, abuse, and exploitation. In the late nineteenth century, after the first case of child abuse was successfully prosecuted, it added child welfare work to its mission and established the American Association for Protecting Children. Activities include training programs for animal care and control professionals, humane education materials and methods for children and adults, emergency relief for animal victims of natural disasters, advocacy for humane legislation and safeguards for animal actors, and public policy to protect abused and neglected families. Membership is $15 and includes the quarterly magazine *Advocate.*

**American Littoral Society** (Sandy Hook, Highlands, NJ 07732; 201-291-0055) is a nonprofit conservation organization founded to study and conserve the littoral zone, the fragile, productive areas where the sea meets the shore. Many local chapters sponsor international field trips, beach cleanups, tern nest patrols, and turtle nest watches. Annual dues are $20 and include the quarterly newsletter *Underwater Naturalist* and the bimonthly newsletter *Coastal Reporter.*

**American Rivers** (801 Pennsylvania Ave. SE, Washington, DC 20003; 202-547-6900) is a nonprofit group working to preserve the nation's rivers and their landscapes. American Rivers measures its progress in river miles preserved, streamside acres protected, dams blocked, and taxpayer dollars saved. Membership dues are $20 and up annually. Members receive the quarterly newsletter *American Rivers,* along with a list of outfitters that support river conservation.

**American Society for Environmental History** (Center for Technology Studies, New Jersey Institute of Technology, Newark, NJ 07012; 201-596-3334) promotes research, publications, teaching, and communications on the relationship of humans to the natural environment from a broadly historical and humanistic perspective. Membership ($24 individuals, $30 institutions) includes the quarterly journal *Environmental Review.*

**American Society for the Prevention of Cruelty to Animals** (441 E. 92nd St., New York, NY 10128; 212-876-7700) was the first humane society in America. Its purpose is to provide effective means for the prevention of cruelty to animals throughout the United States. The group focuses on the rights of companion animals, animals in research and testing, animals raised for food, wild animals, entertainment and work animals, and animals in education. It also sponsors a yearly safari to Kenya to observe wildlife in a natural setting. Membership ($20 per year) includes the *ASPCA Quarterly Report* magazine and a discount on pet products in its New York store.

**American Wilderness Alliance** (7600 E. Arapahoe, Suite 114, Englewood, CO 80112; 303-771-0380) was founded to protect and wisely manage wilderness, wildlife, wetlands, watersheds, fisheries, and quality outdoor recreation. Projects have included reintroduction of beavers to natural habitats, timber management, and water laws. Membership ($22 and up) includes the quarterly publication *It's Time to Go Wild.*

**Americans for Safe Food** (see **Center for Science in the Public Interest**)

**Americans for the Environment** (1400 16th St. NW, Washington, DC 20036; 202-797-6665) is an educational organization established to train, educate, and involve the environmental community in issues and strategy, and to encourage activists to become involved in campaigns to elect environmentalists to public office. This is not a membership organization.

**Animal Protection Institute of America** (2831 Fruitridge Rd., Sacramento, CA 95822; 916-422-1921) was established to eliminate or alleviate fear, pain, and suffering among animals through humane education and member action. Membership ($20 per year) includes a subscription to the quarterly magazine *Mainstream.*

**Animal Welfare Institute** (P.O. Box 3650, Washington, DC 20007; 202-337-2332) promotes the welfare of all animals and works to reduce the pain and fear inflicted on animals by humans. One focus is on improving conditions of laboratory animals, "factory farmed" animals, and species threatened by extinction. It maintains an extensive collection of publications and films for educational use. Membership ($15 annually) includes *Animal Welfare Institute,* a quarterly newsletter, along with options on free books when they become available and invitations to events at the Institute.

**Atlantic Center for the Environment** (39 S. Main St., Ipswich, MA 01938; 508-356-0038) was established to promote environmental understanding and encourage public involvement in resolving natural resource issues in the Atlantic Ocean region. The center has worked on studying migratory birds, river, and watershed management, and various international programs. Annual dues ($25 and up) include a subscription to the organization's quarterly magazine *Nexus*.

**Balloon Alert Project** (12 Pine Fork Dr., Toms River, NJ 08755; 201-341-9506) is an information source for individuals, groups, and schools interested in becoming actively involved in banning mass balloon releases and developing alternatives to such releases. A quarterly newsletter, *Balloon Alert Project Newsletter*, is available by sending a self-addressed, stamped envelope.

**Basic Foundation** (P.O. Box 47012, Saint Petersburg, FL 33743; 813-526-9562) was established to promote efforts to balance population growth with natural resources and tropical rain forest preservation. The foundation supports research activities, exhibits, publications, conferences, lectures, and nature tours. The foundation also publishes and donates educational materials to schools; lobbies policy-makers and international organizations on behalf of the tropical rain forests; sells products with environmental messages; supports activities of various rain forest projects; and sponsors nature tours to rain forests in Costa Rica.

**Bat Conservation International** (P.O. Box 162603, Austin, TX 78716; 512-327-9721) educates about the vital role of bats in world environments. Bats are responsible for controlling insects, pollinating plants, and generating up to 95 percent of the seed dispersal essential to the regrowth of cleared tropical forests. Membership is $25 and up annually and includes the quarterly newsletter *Bats* and a copy of the book *America's Neighborhood Bats*.

**Bio-Integral Resource Center** (P.O. Box 7414, Berkeley, CA 94707; 415-524-2567) is dedicated to providing information on least-toxic pest control. Membership fees depend on subscriptions desired: the *IPM Practitioner*, published ten times per year, is $25; the *Common Sense Pest Control Quarterly* is $30; both are available for $45. Members can receive help with pest-management problems.

**Caribbean Conservation Corp.** (P.O. Box 2866, Gainesville, FL 32602; 904-373-6441) focuses on marine and sea turtle research and conservation. It operates the Green Turtle Research Station in Tortuguero, Costa Rica, and maintains a green turtle tagging project in cooperation with the Center for Sea Turtle Research at the University of Florida. It also runs the Volunteer Research Travel Program, with spring and summer trips to Tortuguero; participants assist research teams in studying and tagging marine turtles. Membership is $35 a year and includes the quarterly newsletter *The Velador*.

**Center for Environmental Information** (99 Court St., Rochester, NY 14604; 716-546-3796) is an information organization that maintains an extensive library, organizes conferences and seminars, and publishes many useful books, manuals, and directories. The center also provides a discount travel club for members. Membership ($25 and up per year) includes a subscription to the bimonthly *Upstate Environment* newspaper, a monthly update of the center's activities (*Sphere*), invitations to monthly "timely topic" seminars, and other publications.

**Center for Investigative Reporting** (530 Howard St., 2nd Floor, San Francisco, CA 94105; 415-543-1200) provides support for investigative journalism, including environmental stories, for international television networks, newspapers, and magazines. The center has a regional office in Washington, D.C. Environmental projects include a documentary for public television focusing on using the earth as a dumping ground for toxic wastes.

**Center for Marine Conservation** (1725 DeSales St. NW, Suite 500, Washington, DC 20036; 202-429-5609), formerly the Center for Environmental Education, is dedicated to protecting marine wildlife and their habitats and conserving coastal and ocean resources. The center conducts policy-oriented research, public awareness through education, and supports international and domestic marine conservation programs. It sponsors the **Whale Protection Fund**, **Marine Habitat Program**, **Seal Rescue Fund**, **Sea Turtle Rescue Fund**, and **Entanglement Program**. Membership ($20 per year) includes the quarterly newsletter *Marine Conservation News*, legislative updates, and "Action Alerts" outlining things people can do to support marine conservation.

**Center for Plant Conservation** (125 Arborway, Jamaica Plain, MA 02130; 617-524-6988) is a national network of nineteen botanical gardens and arboretums. The organization works to conserve rare and endangered native plants through research, cultivation, and education at botanical gardens and arboretums throughout the U.S. The center publishes a newsletter.

**Center for Science in the Public Interest** (1501 16th St. NW, Washington, DC 20036; 202-332-9110) provides consumers with information in the areas of nutrition and health. Research, education, and publication efforts include nutrition advocacy, alcohol and minority-related projects, a child nutrition project, and the **Americans for Safe Food** project, which focuses on sustainable agriculture. Membership is $19.95 per year, to receive *Nutrition Action Health Letter*, published ten times annually.

**Children of the Green Earth** (P.O. Box 95219, Seattle, WA 98145; 206-781-0852) is committed to "regreening" the earth by helping young people plant and care for trees and forests. It publishes educational materials, a periodical newsletter, *Tree Song*, and promotes tree-planting efforts by young people. A partnership program involves groups in Germany, India, Lesotho, South Africa, Senegal, and Nepal. The organization helps others in becoming involved through work in their local communities. Membership is $25 and includes a subscription to the newsletter, as well as a discount on materials.

**Citizen's Clearinghouse for Hazardous Waste** (P.O. Box 3541, Arlington, VA 22216; 703-276-7070) was founded in 1981 by Love Canal victim Lois Gibbs to use the lessons learned at the Love Canal toxic-waste dump in helping grassroots groups fight for environmental justice. It currently works with over 6,000 community groups nationwide, providing technical support for environmental problems. The group publishes many publications geared toward helping communities help themselves—manuals on organizing, fundraising, waste disposal management, how to fight a proposed facility, and how to start a community recycling project. Membership is $25 per year and includes the bimonthly magazine *Everyone's Backyard*.

**Clean Sites** (1199 N. Fairfax St., Alexandria, VA 22314; 703-683-8522) encourages hazardous-waste cleanups conducted by those responsible for the contamination and provides technical reviews and project management services at sites. Clean Sites is supported by reimbursement for services, contributions from the chemical industry, and contributions and grants from corporations, foundations, and government.

**Clean Water Action Project** (317 Pennsylvania Ave. SE, Washington, DC 20003; 202-547-1196) works for clean and safe water at an affordable cost, control of toxic chemicals, and the protection of natural resources. Emphasis is on pesticide safety and groundwater protection, solving the landfill crisis, and protecting endangered natural resources. Membership is $24 for individuals, $40 for organizations. Members receive the quarterly *Clean Water Action News* and monthly regional newsletters.

**Clean Water Fund** (317 Pennsylvania Ave. SE, 3rd Floor, Washington, DC 20005; 202-547-2312) aims to advance environmental and consumer protections and develop the grassroots strength of the environmental movement. The group focuses on water pollution, toxic hazards, and natural resources. Regular membership ($25) includes the quarterly newsletter *Water Action News;* sustaining membership ($60) includes monthly bulletins as well as the newsletter.

**Climate Institute** (316 Pennsylvania Ave. SE, Suite 403, Washington, DC 20003; 202-547-0104) serves as a bridge between scientists and public and private decision-makers on global warming and stratospheric ozone depletion. The organization publishes many reports and proceedings. Membership is $35 ($15 for students) and includes a quarterly newsletter, *Climate Alert.*

**Climate Protection Network** (see **Earth Island Institute**)

**Coalition for Scenic Beauty** (216 7th St. SE, Washington, DC 20003; 202-546-1100) is dedicated to protecting scenic resources in the U.S. and cleaning up visual pollution. The group has worked on billboard control, preservation of scenic areas, and aesthetic regulation. Membership ($20 individual, $50 organization) is open and includes a subscription to *Sign Control News*, a bimonthly newsletter.

**The Coastal Society** (5410 Grosvenor Ln., Suite 110, Bethesda, MD 20814; 301-897-8616) is committed to promoting the understanding and wise use of coastal environments. It sponsors conferences, workshops, and publications. Membership ($25, $12.50 students) includes a quarterly bulletin.

**Coastal States Organization** (444 N. Capitol St. NW, Suite 312, Washington, DC 20001; 202-628-9636) represents the coastal areas (the U.S. has a 95,000-mile coastline) in ocean and coastal affairs. The membership of congressional representatives from those states continually reviews and assesses coastal management practice and policy, problems, and progress throughout the country. It offers a broad-based information and data gathering network, provides information on coastal and offshore development, public access information, coastal hazards planning and management, wetlands preservation, fisheries development and management, and port and waterfront restoration. The group sponsors "Coastweeks" each fall, a three-week volunteer project to clean up the nation's coasts and beaches.

**Committee for Humane Legislation** (see **Friends of Animals**)

**Concern, Inc.** (1794 Columbia Rd. NW, Washington, DC 20009; 202-328-8160) provides environmental information for community action. Concern's Community Outreach program promotes local and regional citizen action and encourages communication between individuals and groups working on similar issues. Its goal is to help communities find solutions to environmental problems that threaten public health and the quality of life. Booklets on pesticide use, farmland, drinking water, groundwater, and household waste are available at $3 each.

**Conservation Foundation** (1250 24th St. NW, Washington, DC 20037; 202-293-4800) is committed to improving the quality of the environment and to securing wise use of the earth's resources by influencing public policy on all levels. The group focuses on pollution and toxic substances, public and private land use in the

U.S., and economic development in the third world. In 1985, the group formally affiliated with the World Wildlife Fund to add a strong scientific background to its activities. The foundation publishes a monthly newsletter for other environmental organizations, the *CF Newsletter*.

**Consumer Pesticide Project** (425 Mississippi St., San Francisco, CA 94105; 415-826-6314) is composed of consumer and environmental activists working to get dangerous pesticides out of food and the environment. The group is dedicated to encouraging citizen participation and focuses on encouraging supermarkets to carry fresh fruits and vegetables without dangerous pesticides. The project's *Organizing Kit: A Practical Strategy to Reduce Dangerous Pesticides in our Food and the Environment* is a step-by-step instruction guide for organizing your community and is available for $5 to cover postage and handling.

**Co-Op America** (2100 M St. NW, Suite 310, Washington, DC 20036; 202-872-5307) is a democratically controlled, nonprofit membership association representing the social and environmental interests of consumers. It believes that to solve the environmental crisis, America must change the way it does business. The group's travel service, Travelinks, arranges ecologically friendly, educational travel. Other benefits include insurance and a credit union that offers a special Visa card. Membership ($20 per year) includes the quarterly magazine *Building Economic Alternatives*.

**Council on Economic Priorities** (30 Irving Pl., New York, NY 10003; 212-420-1133) is a nonprofit research organization devoted to impartial analysis of crucial public-interest issues, including corporate social responsibility, the environment, and national security. Membership ($25 a year and up) includes a copy of the guide *Shopping for a Better World* and a monthly research report.

**Cousteau Society** (930 W. 21st St., Norfolk, VA 23517; 804-627-1144) was founded by world-renowned environmentalist and underwater explorer Jacques-Yves Cousteau to protect and improve the quality of life for present and future generations. The society's activities range from research, lectures, books, and publications to television specials on human interaction with ecosystems. Membership ($20 for individuals, $28 for families) includes the monthly magazine *Calypso Log* (children receive *Dolphin Log*).

**Cultural Survival** (11 Divinity Ave., Cambridge, MA 02138; 617-495-2562) is a Harvard-affiliated academic organization working to import sustainably managed rain forest products to the U.S. It acts as a consultant to businesses that want to import rain forest nuts and woods for use in their products. Cultural Survival manages "Cultural Survival Imports," a nonprofit business importing cashews and Brazil nuts to the U.S. Membership ($25 annually) includes a subscription to *Cultural Survival Quarterly*.

**Defenders of Wildlife** (1244 19th St. NW, Washington, DC 20036; 202-659-9510) is dedicated to protecting wild animals and plants in their natural communities, especially native American endangered or threatened species, through education, litigation, and advocacy of public policies. The group sponsors the Entanglement Network Coalition to identify problems with marine entanglement and debris ingestion. Membership ($20 per year) includes the bimonthly magazine *Defenders*, voting privileges for the Board of Directors, and eligibility for the organization's Visa or MasterCard. It also publishes annual endangered species reports, educational newsletters, and citizen action alerts.

**Desert Fishes Council** (407 W. Line St., Bishop, CA 93514; 619-872-1171) is concerned with the integrity of aquatic ecosystems in the desert Southwest. It supports the research of related agencies and academia. Membership ($10 per year) includes the proceedings of the annual symposium.

**Desert Tortoise Council** (5319 Cerritos Ave., Long Beach, CA 90805; 213-422-6172) was established to assure continued survival of the desert tortoise population. The group advises desert tortoise preservation agencies, fish and wildlife agencies, and any other agencies involved in protection of the desert tortoise and conservation. Membership ($8 individual, $5 student, $20 contributor, $25 organization) is open and includes a quarterly newsletter and notices of symposiums.

**Ducks Unlimited** (1 Waterfowl Way, Long Grove, IL 60047; 312-438-4300) raises money for developing, preserving, restoring, and maintaining the waterfowl habitat in North America. The group promotes public education about wetlands and waterfowl management, and supports the North American Waterfowl Management Plan. Membership is $20 to $10,000 and entitles members to a subscription to the monthly magazine *Ducks Unlimited.*

**Earth First!** (P.O. Box 7, Canton, NY 13617; 315-379-9940) is a radical direct-action movement encouraging individuals to act upon their environmental concerns. The *Earth First! Journal*, published eight times a year, is available for $20 per year.

**Earth Island Institute** (300 Broadway, Suite 28, San Francisco, CA 94133; 415-788-3666) was established to initiate and support internationally oriented projects protecting and restoring the environment. Among Earth Island's projects are the International Marine Mammal Project, the Environmental Project on Central America, the International Rivers Network, and the Climate Protection Network. Membership is $25 annually and includes the quarterly *Earth Island Journal.*

**EarthSave** (P.O. Box 949, Felton, CA 95018; 408-423-4069) developed from the work of John Robbins, author of *Diet for a New America.* It provides education and leadership for transition to more healthful and environmentally sound food choices, nonpolluting energy supplies, and a wiser use of natural resources, and it is dedicated to an ecologically sustainable future. Membership ($20 to $35 per year) includes the *Project EarthSave* newsletter and regular notices and updates of current activities.

**Elsa Wild Animal Appeal** (a.k.a. Elsa Clubs of America, P.O. Box 4572, N. Hollywood, CA 90607; 818-761-8387) is dedicated to the conservation of wildlife, especially endangered species, through child education. It develops and distributes child education materials (for youth under 18) and is involved in legislation. Membership ($15 for adults, $7.50 for youth, $5 for senior citizens) includes a triannual subscription to *Born Free News* and includes a choice of several wildlife kits that can be used at home or in classrooms. Youth members also receive "Action Alerts" throughout the year with titles such as "Elephant Ivory Bans" and "Save the Whales," with suggestions and information about what they can do.

**Energy Conservation Coalition** (see **Environmental Action**)

**Environmental Action Coalition** (625 Broadway, New York, NY 10012; 212-677-1601) promotes recycling in New York City, organizing apartment buildings and urban forestry and educational programs. Membership ($15 to 25 annually) includes the quarterly newsletter *Cycle*.

**Environmental Action, Inc.** (1525 New Hampshire Ave. NW, Washington, DC 20036; 202-745-4870) is a membership-based organization that lobbies Congress for passage of strong environmental laws, such as the Clean Air Act and Superfund. The group works directly with citizen groups on such issues as recycling, right-to-know laws, and toxic pollution. It also sponsors the Solid Waste Alternative Project. The Energy Conservation Coalition branch works through national public interest groups to promote energy efficiency as a solution to such problems as global warming. *Environmental Action Magazine*, published bimonthly, provides in-depth articles on such topics as solid waste, plastic containers, and ecotourism. Membership ($20 per year) includes the magazine.

**Environmental and Energy Study Institute** (122 C St. NW, Suite 700, Washington, DC 20001; 202-628-1400) is a public policy organization aimed at producing more informed congressional debate, credible analysis, and innovative policies for environmentally sustainable development. Programs and projects focus on global climate change, acid rain, groundwater protection, agriculture, solid and hazardous waste management, energy efficiency, and natural resource management in the third world. The group publishes special reports and the *Weekly Bulletin*, published while Congress is in session, which contains highlights of the upcoming week's floor activity in Congress, status reports and forecasts, and positions of key players.

**Environmental Defense Fund** (257 Park Ave. S., New York, NY 10010; 212-505-2100) was founded in 1967 as an organization of scientists, economists, and lawyers defending the environment. It focuses on water pollution, pesticides, wildlife preservation, wetland protection, rain forests, the ozone layer, acid rain, and toxic chemicals and waste. Membership is $20 per year and includes the quarterly *EDF Newsletter*.

**Environmental Hazards Management Institute** (EHMI, 10 Newmarket Rd., P.O. Box 932, Durham, NH 03824; 603-868-1496) aims to educate public and private sector individuals and organizations on hazardous waste, acting as an information and training source. EHMI distributes the "Household Hazardous Waste Whee™l," the "Water Sense Wheel™," and the "Recycling Whee™l". They are available for $3.75 each plus postage and handling, with discounts for quantity orders.

**Environmental Law Institute** (1616 P St. NW, Suite 200, Washington, DC 20036; 202-328-5150) is an environmental law research and education center that helps find creative solutions to such problems as wetlands protection, surface mining, hazardous waste, acid rain, and global warming. The group focuses its efforts on education through courses, publications, technical assistance, and research and policy analysis.

**Environmental Safety** (733 15th St. NW, #1120, Washington, DC 20005; 202-628-0374) is an association of environmental professionals who develop new approaches and alternatives for environmental policy, especially environmental safety. It monitors the activities of the EPA, including budget, personnel, and policy matters, and researches alternatives to EPA policy. It also monitors implementation of the nation's toxic-substance laws and provides information and education materials to the public.

**The Forest Trust** (P.O. Box 9238, Santa Fe, NM 87504; 505-983-8992) was established to protect and improve forest ecosystems and resources. The group's activities focus on national forest management, land trust, management of private lands, and economic development in rural communities.

**Freshwater Foundation** (2500 Shadywood Rd., Box 90, Navarre, MN 55392; 612-471-8407) is dedicated to researching and educating to keep waters usable for human consumption, industry, and recreation. Activities emphasize agricultural chemicals and groundwater protection, and biological processes that degrade pollutants. It publishes *Health and Environment Digest*, *U.S. Water News*, both monthlies, and the biennial *Journal of Freshwater*. Membership is $50 per year.

**Friends of Animals** (P.O. Box 1244, Norwalk, CT 06856; 203-866-5223) is dedicated to eliminating human brutality to animals. Its many programs include breeding-control services, working for the protection of animals used in experiments and testing and for farm animals, and a wild animal orphanage and rehabilitation center in Liberia. It also heads the Committee for Humane Legislation, which is active in legislative affairs. Membership is $20 per year.

**Friends of the Earth** (530 7th St. SE, Washington, DC 20003; 202-544-2600) promotes the conservation, protection, and rational use of the earth. Its activities include lobbying, litigation, and public information on a variety of environmental issues, including ozone depletion, river protection, and tropical deforestation. It recently merged with the Oceanic Society and continues to strives to protect the oceans through education, research, and conservation, and by promoting the understanding and stewardship of the marine and coastal environment. Membership ($25 individual, $15 student/low income/senior citizen) includes the monthly magazine *Not Man Apart*.

**Friends of the River** (Fort Mason Center,Bldg. C, San Francisco, CA 94123; 415-771-0400) was established to save rivers from being dammed. It is active in river conservation efforts, and has successfully fought for preservation of the Tuolumne, Kings, Kern, and Merced rivers in California. Membership is $25 for individuals, $35 for families; members receive the bimonthly *Headwaters* as well as "action alerts" and discounts on river trips.

**Friends of the Sea Otter** (Box 221220, Carmel, CA 93922; 408-625-3290) aids in the protection and maintenance of the southern sea otter and its near-shore marine habitats. Membership is $15 per year and includes the twice-annual *Otter Raft* magazine.

**Fund for Animals** (200 W. 57th St., New York, NY 10019; 212-246-2096) was founded to aid the relief of fear, pain, and suffering in wild and domestic animals. Annual membership is $10 for students, $20 for individuals, and $25 for families. Members receive a newsletter and "Action Alert" updates on legislation.

**Global Tomorrow Coalition** (1325 G St. NW, Washington, DC 20005; 202-628-4016) is a national alliance of organizations and individuals created to foster broader understanding of the long-term significance of global trends in population, resources, environment, and development. Its main focus is on sustainable development. The group also helps promote informed and responsible public choice among alternative futures for the U.S. and alternative roles for the nation within the international community. Membership is $15 a year for senior citizens and students and $35 for individuals and includes the quarterly newsletter *InterAction*.

**Grass Roots the Organic Way** (38 Llangollen Lane, Newton Square, PA 19073; 215-353-2838) provides information about harmful pesticides and safe alternatives. It educates on the use and misuse of pesticides on lawns, trees, shrubs, and other indoor and outdoor plants. Membership ($20) entitles members to call with specific problems. The group will either provide solutions over the phone or will send an information packet on the problem.

**Great Bear Foundation** (P.O. Box 2699, Missoula, MT 59806; 406-721-3009) is dedicated to conservation of bears, especially grizzly bears, and their habitat through such means as public education and habitat monitoring. Membership is $15 for individuals, $25 for families, and includes the quarterly newsletter *Bear News*.

**Great Swamp Research Institute** (Office of the Dean, College of Natural Sciences and Math, Indiana University of Pennsylvania, 305 Weyandt Hall, Indiana, PA 15705; 412-357-2609) is a research and educational organization committed to protecting, preserving, and maintaining the environment. It seeks new solutions to deal with the daily pressures society is placing on the environment. Membership is free; write for information.

**Greater Yellowstone Coalition** (P.O. Box 1874, Bozeman, MT 59715; 406-586-1593) was established to preserve and protect the wildlife, wildlands, fisheries, and other natural wonders in and around Yellowstone National Park. Membership is $20 for individuals, $50 for organizations, per year.

**Greenhouse Crisis Foundation** (1130 17th St. NW, Suite 630, Washington, DC 20036; 202-466-2823), a project of the Foundation on Economic Trends, is dedicated to creating global awareness of the greenhouse crisis and to changing the world view and lifestyle underlying that crisis. It has initiated several local, national, and international actions, including the Greenhouse Education Campaign; the Cities Program to raise awareness of mayors throughout the world of the importance of energy conservation, urban reforestation, and mass transit; International Environmental Rights Conferences; and the Global Greenhouse Network, to raise awareness and mobilize public opinion and to help facilitate international cooperation.

**Greenpeace** (1436 U St. NW, Washington, DC 20009; 202-462-1177) is dedicated to protecting and preserving the environment and the life it supports. It has focused its efforts on halting the needless killing of marine mammals and other endangered animals, ocean ecology, toxic waste reduction, and nuclear disarmament. Membership ($20 per year) includes the bimonthly *Greenpeace* magazine.

**Household Hazardous Waste Project** (901 S. National Ave., Box 108, Springfield, MO 65804; 417-836-5777) educates the public on household hazardous waste. It provides consumer information, offers training and materials to community groups, and supports a grassroots approach to working on household hazardous waste issues.

**Human Ecology Action League** (P.O. Box 49126, Atlanta, GA 30359; 404-248-1898) aims to increase awareness of environmental conditions that are hazardous to human health. The group acts as a clearinghouse on chemical sensitivities and related disorders, works toward minimizing the indiscriminate use of harmful chemicals, and establishes local chapters that provide support for members and educate their communities. Membership ($20) includes a subscription to the newsletter *The Human Ecologist*.

**Human Environment Center** (1001 Connecticut Ave. NW, Suite 827, Washington, DC 20006; 202-331-8387) is dedicated to providing education, information, and services to encourage the integration of environmental organizations, and promoting joint activities among environmental and social equity groups. It serves as a clearinghouse and technical assistance center for youth conservation and service corps programs and operates a recruitment and placement service for minority environmental interns and professionals.

**Humane Society of the United States** (2100 L St. NW, Washington, DC 20037; 202-452-1100) offers resources to the general public on such topics as animal control, cruelty investigation, publications, and humane education. The group's efforts include a "Shame of Fur" campaign, "Be a P.A.L.—Prevent A Litter" campaign, marine mammal protection, and laboratory animal welfare. Membership is $10 per year.

**Infact** (256 Hanover, Boston, MA 02113; 617-742-4583) was established to press for corporate accountability and responsibility. The group has research offices in several cities and national campaign headquarters in California. One focus is on protesting the nuclear weapons efforts of General Electric. The publication *Infact Brings GE to Light* is available for $8.45. Membership is $15 a year.

**Inform** (381 Park Ave. S., New York, NY 10016; 212-689-4040) was established to conduct environmental research and education on such topics as garbage management, industrial toxic waste reduction, urban air pollution, and land and water conservation. Membership ($35 and up) includes a bimonthly newsletter.

**Institute for Earth Education** (Box 288, Warrenville, IL 60555; 312-393-3096) is dedicated to developing a serious educational response to the environmental crisis of the earth. It consists of a network of individuals and member organizations committed to fostering earth education programs. The group conducts workshops, publishes a quarterly journal, *Talking Leaves*, hosts international and regional conferences, supports local branches, and publishes books. Membership ($20 and up) includes a subscription to the journal. A free *Sourcebook* outlining the group's Earth Education program materials is available upon request.

**Institute for Local Self-Reliance** (2425 18th St. NW, Washington, DC 20009; 202-232-4108) helps communities achieve maximum use of their physical, financial, and human resources. Its activities include research, educational workshops, and direct consulting services to community groups and local governments. The group can also provide data bases for research.

**International Alliance for Sustainable Agriculture** (Newman Center, University of Minnesota, 1701 University Ave. SE, Room 202, Minneapolis, MN 55414; 612-331-1099) works to promote sustainable agriculture worldwide in an economically viable, ecologically sound, socially just, and humane way. The group focuses on three problems in sustainable agriculture: insufficient research and documentation; lack of organizational support and network building; and inadequate education and information dissemination. Membership ($10 to $1,000) includes a subscription to *Manna*, a bimonthly newsletter, and discounts on other publications.

**International Council for Bird Preservation** (801 Pennsylvania Ave. SE, Washington, DC 20003; 202-778-9563) was established to help maintain the diversity, distribution, abundance, and natural habitats of bird species worldwide, and to prevent the extinction of any species or subspecies. Activities have focused on monitoring the status of susceptible bird populations and fostering international cooperation in bird preservation efforts. Membership ($35 and up) entitles members to two quarterly magazines, *World Bird Watch* and *U.S. Bird News*.

**International Fund for Animal Welfare** (P.O. Box 193, 411 Main St., Yarmouth Port, MA 02675; 508-362-4944) is dedicated to the protection of wild and domestic animals. Efforts have included work to preserve harp and hood seals in Canada, fur seals in Alaska, and vicuna in Peru. Membership is by donation; benefits include a variety of newsletters.

**International Marine Mammal Project** (see **Earth Island Institute**)

**International Oceanographic Foundation** (4600 Rickenbacker Causeway, P.O. Box 499900, Miami, FL 33149; 305-361-4888) provides information about the world's oceans and their importance to humanity, and encourages scientific investigation of the ocean. It operates a museum at its Miami headquarters. Membership is $18 and entitles members to receive the bimonthly magazine *Sea Frontiers*.

**International Primate Protection League** (P.O. Box 766, Sumerville, SC 28484; 803-871-2280) is dedicated to the conservation and protection of primates. Activities focus on primate trafficking, laboratory primate issues, and maintenance of a gibbon sanctuary. Membership is $20 and includes the quarterly newsletter *International Primate Protection League*.

**International Rivers Network** (see **Earth Island Institute**)

**International Society of Arboriculture** (303 W. University Ave., Urbana, IL 61801; 217-328-2032) is dedicated to proper tree care and preservation, particularly in urban settings. Its activities are aimed at helping make the public aware of the impact trees have on our future. The group provides services to its members about the science and art of growing and maintaining shade and landscape trees. Membership ($55 per year) includes a subscription to the monthly *Journal of Arboriculture*.

**International Society of Tropical Foresters** (5400 Grosvenor Lane, Bethesda, MD 20814; 301-897-8720) is dedicated to protecting, wisely managing, and rationally using the world's tropical forests. Activities include establishing a communications network among tropical foresters and others concerned with the forests. Membership ($50 annually) includes a subscription to a quarterly newsletter *ISTF News*, and the monthly magazine *Journal of Forestry*.

**Izaak Walton League of America** (1701 N. Fort Myer Dr., Arlington, VA 22209; 703-528-1818) aims to protect America's land, water, and air resources. Efforts focus on acid rain, clean air and water, stream protection, soil erosion, the Chesapeake Bay cleanup, and waterfowl/wildlife protection. The organization coordinates the programs Save Our Streams and Wetlands Watch. Membership ($20 per year) includes a quarterly magazine *Outdoor America.*

**League of Conservation Voters** (320 4th St. NE, Washington, DC 20002; 202-785-8683) is a national, nonpartisan political arm of the environmental movement. It works to elect pro-environmental candidates to Congress, based on energy, environment, and natural-resource issues. Membership is $25 per year and includes *The National Environmental Scorecard*, an annual rating of members of Congress on environmental issues.

**League of Women Voters of the United States** (1730 M St. NW, Washington, DC 20036; 202-429-1965) is a nonpartisan, political organization that encourages informed, active participation of citizens in government and influences public policy through education and advocacy. The league takes political action on water and air quality, solid- and hazardous-waste management, land use, and energy. Membership is $50 per year for a national membership (local memberships vary) and includes a subscription to the monthly *National Voter* magazine.

**Marine Mammal Stranding Center** (P.O. Box 733, Brigantine, NJ 08203; 609-266-0538) is a rescue and rehabilitation organization for marine mammals and sea turtles. It sponsors many whale-, dolphin-, and seal-watching trips. Membership is $10 per year and includes the quarterly newsletter *Blow Hole.*

**Monitor Consortium of Conservation and Animal Welfare Organizations** (1506 19th St. NW, Washington, DC 20036; 202-234-6576) was founded in 1972 as a nonprofit coordinating center and information clearinghouse on endangered species and marine mammals for its member organizations. Membership is open to conservation, environmental, and animal welfare organizations.

**National Arbor Day Foundation** (100 Arbor Ave., Nebraska City, NE 68410; 402-474-5655) is dedicated to tree planting and conservation. It provides direction, technical assistance, and public recognition for urban and community forestry programs. Projects include Tree City USA, an urban forestry tree planting and care program; Friends of Tree City, for individuals in urban areas that are interested in trees; Trees for America; and Conservation Trees. The group provides materials and information for cities to plan their own Arbor Day celebrations, including children's packets. Membership ($10 per year) includes the bimonthly newsletter *Arbor Day* and a tree book.

**National Audubon Society** (950 Third Ave., New York, NY 10022; 212-832-3200) aims to conserve native plants and animals and their habitats; protect life from pollution, radiation, and toxic substances; further the wise use of land and water; seek solutions for global problems involving the interaction of population, resources, and the environment; and promote rational strategies for renewable energy development. Membership ($30 per year) includes a subscription to the bimonthly *Audubon Magazine*. A major activity is the 12,000-member Activist Network. Network members volunteer to act upon their environmental concerns, especially in times of environmental crisis, such as an oil spill. Membership in the network ($9 per year) includes a subscription to the bimonthly *Audubon Activist* and *Action Alerts*, one-page notices reporting the status of environmental legislation. The society also offers environmentally responsible travel programs, elementary school education programs, adult conservation camps, and an ornithological journal, *American Birds*.

**National Clean Air Coalition** (530 7th St. SE, Washington, DC 20003; 202-543-8200) is a lobbying and education coalition that aims to address clean-air issues. Membership is free and includes a periodical newsletter, *Clean Air 101*.

**National Coalition Against the Misuse of Pesticides** (530 7th St. SE, Washington, DC 20003; 202-543-5450) assists individuals, organizations, and communities with information on pesticides and their alternatives. The group maintains an information clearinghouse that provides materials on agricultural and urban issues concerning lawn-care safety, farm workers' safety, groundwater problems, and alternatives to pesticides. Membership ($20) includes a subscription to the newsletter *Pesticides and You*.

**National Coalition for Marine Conservation** (P.O. Box 23298, Savannah, GA 31403; 912-234-8062) is dedicated to conserving oceanic ecosystems and the habitat areas that support them. It works to educate policy-makers at all levels of government. Efforts focus on fishery management, oceanic dumping, and wetlands preservation. Membership ($25 annually) includes a bimonthly newsletter *Marine Bulletin*.

**National Geographic Society** (17th and M Streets NW, Washington, DC 20036; 202-857-7000), founded in 1888 "for the increase and diffusion of geographic knowledge," is now the world's largest scientific and educational nonprofit organization, with 10.8 million members. Its main areas of activity are publishing four magazines, books for adults and children, and atlases; producing television programs; supporting environmental research and a geography education program for schoolchildren. Membership is $21 a year and includes a subscription to *National Geographic* magazine.

**National Institute for Urban Wildlife** (10921 Trotting Ridge Way, Columbia, MD 21044; 301-596-3311) is devoted to wildlife research, management, and conservation education programs and activities. Projects have included discovering practical procedures for maintaining or enhancing wildlife species in urban areas. Membership ($25) includes the quarterly newsletter *Urban Wildlife News*.

**National Parks and Conservation Association** (1015 31st St. NW, Washington, DC 20007; 202-944-8530) is dedicated to defending, promoting, and improving America's national park system while educating the public about the parks. Membership ($25) entitles members to the bimonthly magazine *National Parks* as well as discounts on car rentals and camera film.

**National Recycling Coalition** (1101 30th St. NW, Suite 305, Washington, D.C. 20006; 202-625-6406) promotes increased opportunities for recycling. It focuses on education, information, and lobbying, and sponsors the largest annual conference on recycling, the Annual National Recycling Congress. The group's Technical Assistance Program sends advisers to cities that want to establish recycling programs. Membership ($30 per year) includes a bimonthly newsletter *NRC Connection*.

**National Toxics Campaign** (37 Temple Pl., 4th Floor, Boston, MA 02111; 617-482-1477) works on Superfund-related activities. The group provides citizen outreach and educational efforts, political organizing, and a research library. It also maintains a testing lab and legal offices that provide advice to communities dealing with hazardous-waste issues. Membership is $25 for an individual and $50 and up for community groups, and includes the quarterly magazine *Toxic Times*.

**The National Water Center** (P.O. Box 264, Eureka Springs, AR 72632; 501-253-9755) aims to gather, distill, and disseminate information on water issues, emphasizing personal responsibility for human and hazardous waste. Activities include a focus on promoting composting toilets. Membership (minimum $10 per year) includes the newsletter *Water Center News*.

**National Wildlife Federation** (1400 16th St. NW, Washington, DC 20036; 202-797-6800) was founded in 1936 "to be the most responsible and effective conservation education association promoting the wise use of natural resources and protection of the global environment." It distributes periodicals and educational materials, sponsors outdoor education programs in conservation, and litigates environmental disputes. Efforts have focused on forests, energy, toxic pollution, environmental quality, biotechnical fisheries and wildlife, wetlands, water resources, and public lands. Membership rates vary depending on magazine subscriptions desired: *National Wildlife* and *International Wildlife*, monthly magazines, are $15 each per year; *Ranger Rick*, a monthly magazine for children 6 to 12 years old, is $14 per year; *Your Big Backyard*, a monthly magazine for children aged 3 to 5 years old, is $10.

**Natural Resources Defense Council** (40 W. 20th St., New York, NY 10011; 212-727-2700) is committed to protecting America's endangered natural resources and to improving the quality of the environment. It monitors government agencies, brings legal actions, and disseminates information. Areas of focus include air and water pollution, global warming, urban environment, toxic substances control, resources management, energy conservation, and Alaska. Membership ($10) includes a subscription to *The Amicus Journal* (quarterly) and the *Natural Resources Defense Council Newsline*, a bimonthly newsletter.

**The Nature Conservancy** (1815 N. Lynn St., Arlington, VA 22209; 703-841-5300) acts to preserve ecosystems and the rare species and communities they shelter. The group has protected more than 3.5 million acres of threatened habitat, mostly by purchasing land, and manages more than 1,000 preserves. It maintains a database, the Heritage Network, which is a national inventory of species in each state. Membership is $15 and includes a bimonthly magazine *The Nature Conservancy Magazine*, as well as newsletters and update information from state chapters.

**North American Lake Management Society** (P.O. Box 217, Merrifield, VA 22116; 202-466-8550) is dedicated to promoting a better understanding of lakes, ponds, reservoirs, impoundments, and their watersheds. The group encourages the exchange of information about lake management; provides guidance to public and private agencies involved in lake management; and promotes research on lake ecology and watershed management. Membership ($20 and up) includes a periodic newsletter *Lakeline*.

**Oceanic Society** (see **Friends of the Earth**)

**Pacific Whale Foundation** (101 N. Kihei Rd., Suite 21, Kihei, Maui, HI 96753; 808-879-8811, 800-942-5311) is dedicated to the science of saving whales and the ocean environment. It conducts public education programs on a variety of cetacean, marine conservation, and pollution issues, and it sponsors a Marine Debris Cleanup Day and Whale Day. Membership is $15 for students and seniors, $20 for individuals, and $25 for families.

**Pennsylvania Resources Council** (25 W. 3rd St., P.O. Box 88, Media, PA 19063; 215-565-9131) sponsors a statewide recycling conference every spring and publishes *Environmental Shopping*, a booklet designed to help shoppers choose environmentally safe products. The group publishes two newsletters, *PRC News* (3 times per year) and *All About Recycling* (quarterly). Newsletters are free with membership ($30 for individuals; $35 for nonprofit organizations; $50 and up for businesses).

**People for the Ethical Treatment of Animals** (P.O. Box 42516, Washington, DC 20015; 301-770-7444) aims to end the exploitation and abuse of animals, with a special focus on animals used in laboratory experiments. The group sponsors an anti-fur campaign and a vegetarian campaign as well as lobbying, public education efforts, and demonstrations for animal rights. Membership ($15 per year) includes the bimonthly news magazine *PETA News.*

**The Peregrine Fund** (5666 W. Flying Hawk Lane, Boise, ID 83709; 208-362-3716) is devoted to preserving rare and endangered birds of prey worldwide. It coordinates field projects to study birds of prey and conducts reintroduction programs and educational programs. Projects include the Peregrine Falcon Recovery, Aplomado Falcon Recovery, and tropical raptor surveys. Membership is $25 a year and includes *The Peregrine Fund Newsletter.*

**Pesticide Action Network** (P.O. Box 610, San Francisco, CA 94101; 415-541-9140) is a worldwide coalition of more than 300 organizations in more than 50 countries working to stop pesticide misuse and global pesticide proliferation and work toward safe, sustainable pest control. The group produces two triannual newsletters, the *Global Pesticide Monitor* and the *Dirty Dozen Campaigner,* the bimonthly *PANNA Outlook,* and publications on pesticides and alternatives. Membership ranges from $30 to $50 and includes a subscription to *Global Pesticide Monitor.*

**Public Citizen** (2000 P St. NW, Washington, DC 20036; 202-293-9142) was established by Ralph Nader in the interest of consumer protection. It conducts research, lobbying, lawsuits, expert testimony, and publications on issues including stronger environmental programs and safe energy. Membership ($20 per year) includes a subscription to the monthly *Public Citizen Magazine.*

**Public Voice for Food and Health Policy** (1001 Connecticut Ave. NW, Suite 522, Washington, DC 20036; 202-659-5930) is a consumer research, education, and advocacy organization working on issues related to health, nutrition, and food safety. Special emphasis is placed on establishing comprehensive, mandatory seafood testing; access to a safe, affordable, and nutritious food supply; and women's health issues. Membership is $20 per year and includes the quarterly *Action Alert* bulletins highlighting recent legislation.

**Rachel Carson Council** (8940 Jones Mill Rd., Chevy Chase, MD 20815; 301-652-1877) is an international clearinghouse of information on ecology of the environment for all on chemical contamination. It sponsors conferences and produces publications. Subscription to its newsletter is $15 per year and includes Council Publications for one year.

**Rainforest Action Movement** (430 E. University, Ann Arbor, MI 48109; 313-764-2147) focuses on the preservation, protection, and rational use of rain forests in Alaska, Oregon, Washington, Hawaii, and tropical areas. It sponsors public awareness programs in Ann Arbor and surrounding areas, including programs for schoolchildren, lectures, films, citizen forums, benefit concerts, and weekly activities on the campus. Its newsletter *Tropical Echoes*, published every six weeks, is $10 a year.

**Rainforest Action Network** (300 Broadway, Suite 28, San Francisco, CA 94133; 415-398-2732) focuses exclusively on rain forest protection. It works with other environmental and human rights organizations on major rain forest protection campaigns. Topics of emphasis include tropical timber, indigenous peoples, and multilateral development banks. The group has educational publications and materials, a slide show, and press kits available. Membership ($25 regular, $15 limited income) includes two newsletters, *Rain Forest Action Alert* (monthly) and *World Rain Forest Report* (quarterly).

**Rainforest Alliance** (270 Lafayette St., Suite 512, New York, NY 10012; 212-941-1900) aims to link individuals interested in saving tropical rain forests. The group brings together conservation groups, professional organizations, financial institutions, scientists, the business community, and concerned individuals. Membership is $20 per year, $15 for students and senior citizens, and includes a subscription to the quarterly newsletter *Canopy*.

**Renew America** (1001 Connecticut Ave. NW, Suite 719, Washington, DC 20036; 202-232-2252) is an educational and networking forum devoted to the efficient use of natural resources, working at the federal, state, and private citizen levels. Membership ($25) includes the quarterly *Renew America Report* and a copy of the yearly *State of the States* report, which provides an annual "report card" on current developments across the nation.

**Resources for the Future** (1616 P St. NW, Washington, DC 20036; 202-328-5000) conducts research on the environment and the conservation and development of natural resources, including air and water pollution, solid waste disposal, pesticides, toxic substances, and international issues. It publishes various research and technical publications. Although it is not a membership organization, a quarterly newsletter, *Resources,* is available free.

**Rocky Mountain Institute** (1739 Snowmass Creek Rd., Old Snowmass, CO 81654; 303-927-3128) aims to foster the efficient and sustainable use of resources as a path to global security. It offers its research on resource efficiency, global security, and community economic renewal through its publications. A publication list is available upon request.

**Save Our Streams** (see **Izaak Walton League of America**)

**Save the Redwoods League** (114 Sansome St., Room 605, San Francisco, CA 94104; 415-362-2352) was established to rescue areas of primeval forest from destruction. The league purchases redwood groves by private subscription and encourages a better, more general understanding of the value of primeval forests. Projects have included protecting giant sequoias at Sequoia National Forest, working toward establishing the Smith Wild and Scenic National Park in northern California, and purchasing additional redwood lands. Membership ($10 and up) includes spring and fall bulletins on the league's activities.

**Sea Shepherd Conservation Society** (P.O. Box 7000-S, Redondo Beach, CA 90277; 213-373-6979) is an all-volunteer organization established to protect marine animals and marine habitats by direct action. Activities include preventing the killing of dolphins by the tuna industry, protecting pilot whales in the Faroe Islands, and rescuing whales and marine mammals in distress. Membership is by donation. Members receive the quarterly newsletter *Sea Shepherd Log*.

**Sierra Club** (730 Polk St., San Francisco, CA 94109; 415-776-2211) promotes conservation of the natural environment by influencing public policy decisions. It was founded in 1892 to explore, enjoy, and protect the wild places on earth; to practice and promote the responsible use of the earth's ecosystems and resources; and to educate and enlist humanity to protect and restore the quality of the natural and human environment. Campaigns include the Clean Air Act reauthorization, arctic national wildlife refuge protection, national parks and forest protection, global warming/ greenhouse effect, and international development lending reform. Membership ($33 per year) includes a subscription to the monthly magazine *Sierra* and chapter publications. The club also offers more than 275 outings annually. The group's **Legal Defense Fund** (2044 Fillmore St., San Francisco, CA 94115; 415-567-6100) supports lawsuits brought on behalf of citizens' organizations to protect the environment.

**Soil and Water Conservation Society of America** (7515 NE Ankeny Rd., Ankeny, IA 50021; 515-289-2331) is a scientific and educational organization dedicated to conservation of land, water, and other natural resources. Efforts focus on low-input sustainable agriculture, water quality, land and water conservation, and evaluation of farm legislation. The society sponsors a scholarship program for high school and college students. Membership (first year $25, $37 thereafter) includes the *Journal of Soil and Water Conservation*, a bimonthly, and a newsletter covering the annual meeting.

**Student Conservation Association** (P.O. Box 550, Charlestown, NH 03603; 603-826-4301) is an educational organization providing high school and college students and others with the opportunity to volunteer their services for the better management of national parks, forests, public lands, and natural resources. Membership ($10 for students, $25 for others) includes the annual programs listing and the newsletter *The Volunteer*.

**Student Pugwash** (1638 R St. NW, Suite 32, Washington, DC 20009; 202-328-6555) sponsors a variety of educational programs for university students, preparing them as "future leaders and concerned citizens to make thoughtful decisions about the use of technology." Student Pugwash runs international conferences, a "New Careers" program, has chapters on 30 campuses nationwide, and distributes educational products and publications. The group produces several publications, including two newlsetters: *Pugwatch*, distributed to university students, and *Tough Questions*.

**TreePeople** (12601 Mulholland Dr., Beverly Hills, CA 90210; 818-753-4600) is dedicated to promoting personal involvement, community action, and global awareness of environmental issues. Efforts include teaching people how to plant and maintain trees, environmental leadership programs for children, and reforestation efforts in the mountains surrounding Los Angeles. Membership ($25 and up) includes the bimonthly newsletter *Seedling News* and six free seedlings each year.

**Trees for Life** (1103 Jefferson, Wichita, KS 67203; 316-263-7294) provides funding, management, and know-how to people in developing countries to plant and care for food-bearing trees. It runs the "Grow-a-Tree" program, encouraging children to plant trees, and distributes packets of materials, seeds, and instructions to schools and summer camps. Membership (free) includes a subscription to the quarterly newsletter *Life Lines*.

**Trust for Public Land** (116 New Montgomery St., San Francisco, CA 94105; 415-495-4014) aims to conserve land as a living resource for present and future generations. The trust has helped or established more than 150 local land trusts and has acted to preserve nearly 500,000 acres under public ownership. Activities focus on urban waterfronts, suburban greenways, wetlands, agricultural lands, and inner-city open spaces.

**U.S. Public Interest Research Group** (215 Pennsylvania Ave. SE, Washington, DC 20003; 202-546-9707) focuses on consumer and environmental protection, energy policy, and governmental and corporate reform. Efforts include monitoring the implementation of Superfund and legislation for clean air and pesticide safety. Membership ($25) includes the quarterly *Citizen's Agenda*.

**Union of Concerned Scientists** (26 Church St., Cambridge, MA 02138; 617-547-5552), perhaps best known for its "doomsday clock," is a coalition of scientists, engineers, and other professionals concerned with health, safety, environmental, and national security problems posed by nuclear energy and weapons. It conducts policy and technical research, public education, and legislative advocacy on advanced-technology issues.

**Wetlands Watch** (see **Izaak Walton League of America**)

**The Whale Center** (3929 Piedmont Ave., Oakland, CA 94611; 415-654-6621) promotes whales and their ocean habitats through conservation, education, research, and advocacy. Activities have centered on a whaling moratorium, establishment of national marine sanctuaries, the "WhaleBus" mobile classroom, and the "Adopt-a-Gray-Whale" program. Annual membership is $25 for individuals, $10 for students and seniors, $35 for families, and $250 for "life" members. Members receive the center's quarterly newsletter *The Whale Center Journal*.

**The Wilderness Society** (1400 Eye St. NW, Washington, DC 20005; 202-842-3400) aims to protect wildlands, wildlife, forests, parks, rivers, and shorelands. Efforts focus on an Arctic wildlife refuge, national park and ecosystem management, and a national forest policy. Membership ($15 the first year, $30 subsequent years) includes a subscription to the bimonthly newsletter *The Wildlifer*.

**Wildlife Conservation International** (New York Zoological Society, Bronx, NY 10460; 202-367-1010), a division of the New York Zoological Society, is an international conservation program. It conducts field research and conservation action geared toward obtaining a more complete understanding of the biology of endangered species and the structure, functioning, and stability of large ecosystems. Membership ($23 associates, $50 supporting) includes a subscription to the *Wildlife Conservation International Newsletter*.

**Wildlife Information Center** (629 Green St., Allentown, PA 18102; 215-434-1637) is dedicated to securing and disseminating wildlife conservation, recreation, and scientific research information. Its programs include in-service teacher training courses and other public education, maintaining a Wildlife Conservation Registry of Fame, sponsoring wildlife conferences, and advocating nonkilling uses of wildlife such as observation, photography, sound recording, drawing and painting, and wildlife tourism. Projects include securing a ban on importation and sale of live wild birds as pets and banning the use of pole traps. Membership ($25 and up) includes a subscription to *Wildlife Activist*.

**Wildlife Society** (5410 Grosvenor Lane, Bethesda, MD 20814; 301-897-9770) is active in scientific management of the earth's wildlife resources management. Membership ($20 individual, $10 students) includes a subscription to *The Journal of Wildlife Management*, *The Wildlifer*, and *Wildlife Society Bulletin*.

**Windstar Foundation** (2317 Snowmass Creek Rd., Snowmass, CO 81654; 303-927-4777) is an educational organization dedicated to the belief that responsible personal action is the key to creating a sustainable future on a global scale. It conducts research, develops demonstration projects, and offers educational programs. Membership ($35 per year) includes the quarterly *Windstar Journal*.

**World Environment Center** (419 Park Ave. S., New York, NY 10016; 212-683-4700) serves as a bridge between industry and government to strengthen environmental management and industrial safety through exchange of information and technical expertise. The center publishes *The World Environment Handbook*.

**World Resources Institute** (1735 New York Ave. NW, Washington, DC 20006; 202-638-6300) helps the government, the private sector, and organizations address issues in environmental integrity, resources management, economic growth, and international security. Each year it publishes the *World Resources Report*.

**Worldwatch Institute** (1776 Massachusetts Ave. NW, Washington, DC 20036; 202-452-1999) is an independent research organization alerting decision-makers and the general public to emerging global trends in the availability and management of resources. The results of its research are published in Worldwatch papers and books. It also publishes the *State of the World* series, an annual report on the world's resources and their management. Worldwatch papers and books are available by subscription ($25 per year). It also publishes a bimonthly magazine *World Watch* ($20 per year).

**WorldWIDE** (1250 24th St. NW, Washington, DC 20037; 202-331-9863) aims to mobilize women to maintain and improve environmental quality and natural resource management, and to educate the public about the linkages among women, natural resources, and sustainable development. It sponsors WorldWIDE forums and publishes the *Directory of Women in the Environment*. Membership ($35 per year) includes a subscription to the newsletter *WorldWIDE News*.

**World Wildlife Fund** (1250 24th St. NW, Washington, DC 20037; 202-293-4800) works to protect endangered wildlife and wildlands. Its top priority is conservation of the tropical forests in Latin America, Asia, and Africa. The group supports individuals and institutions carrying out practical, scientifically based conservation projects. Members ($15 per year) receive *Focus*, a bimonthly newsletter and periodic letters about upcoming projects and travel programs.

**The Xerces Society** (10 SW Ash St., Portland, OR 97204; 503-222-2788) promotes the global protection of invertebrate habitats, and fosters positive public knowledge of insects by emphasizing their beneficial roles in natural ecosystems. Efforts have aimed at preserving monarch butterfly overwintering habitat and creating a database of invertebrate specialists. Membership ($15 students/seniors/retirees, $25 others) includes a subscription to the magazine *Wings*.

## -Federal Government Environmental Offices-

Below are addresses, phone numbers, and brief descriptions of congressional and administrative agencies of the federal government. If you are not sure where to direct your inquiry, call the nearest Federal Information Center, listed in the government section of your telephone directory. Also helpful is the local office of your House or Senate representatives (their numbers are also in the phone book). Their staffs can direct you to the correct agency or office. Also, be aware that most of the agencies headquartered in Washington, D.C., maintain offices around the country.

### Congress

House Agriculture Committee
U.S. House of Representatives
Washington, DC 20515
202-225-2171
*Jurisdiction over legislation on pesticides.*

House Appropriations Committee
U.S. House of Representatives
Washington, DC 20515
202-225-2771
*Jurisdiction over Environmental Protection Agency, Council on Environmental Quality, National Science Foundation, Marine Mammal Commission, and parts of the Interior Department, Commerce Department, and other agencies.*

House Energy and Commerce Committee
U.S. House of Representatives
Washington, DC 20515
202-225-2927
*Jurisdiction over matters related to solid-waste disposal, pollution, toxic substances, and other hazardous materials.*

House Interior Committee
U.S. House of Representatives
Washington, DC 20515
202-225-2761
*Jurisdiction over many aspects of natural resources.*

House Merchant Marine and Fisheries Committee
U.S. House of Representatives
Washington, DC 20515
202-225-4047
*Jurisdiction over many aspects of natural resources, including conservation.*

House Public Works and Transportation Committee
U.S. House of Representatives
Washington, DC 20515
202-225-4472
*Jurisdiction over all transportation, highway, and mass-transit matters.*

Senate Agriculture Nutrition and Forestry Committee
U.S. Senate
Washington, DC 20510
202-224-2035
*Jurisdiction over many natural resources and environmental issues, including Superfund, the National Forest Service, and parts of the Interior Department.*

Senate Appropriations Committee
U.S. Senate
Washington, DC 20510
202-224-0334
*Jurisdiction over Environmental Protection Agency, Council on Environmental Quality, National Science Foundation, Forest Service, and parts of the Interior Department, Energy Department, Commerce Department, and other agencies.*

Senate Commerce, Science, and Transportation Committee
U.S. Senate
Washington, DC 20510
202-224-5115
*Jurisdiction over ocean-related matters.*

Senate Energy and Natural Resources Committee
U.S. Senate
Washington, DC 20510
202-224-4971
*Jurisdiction over many aspects of natural resources, including research and development.*

Senate Environment and Public Works Committee
U.S. Senate
Washington, DC 20510
202-224-6176
*Jurisdiction over most environmental matters, including pollution control.*

## Federal Agencies

**Department of Agriculture**
Soil Conservation Service
14th St. and Independence Ave. SW
P.O. Box 2890
Washington, DC 20013
202-447-4543
*Oversees all soil and water conservation programs.*

U.S. Forest Service
14th St. and Independence Ave. SW
P.O. Box 96090
Washington, DC 20250
202-447-2791
*Manages all national forests and grasslands, including overseeing policies related to timber harvesting, water, fish, and wildlife.*

**Department of Commerce**
National Marine Fisheries Service
1335 East-West Highway
Silver Spring, MD 20910
301-427-2370
*Responsible for protecting whales, porpoises, seals, sea lions, and other marine mammals, especially endangered species.*

National Oceanic and Atmospheric Administration
14th St. and Constitution Ave. NW
Washington, DC 20230
202-377-8090
*Oversees National Weather Service and protection of coasts and inland waterways, including many conservation activities.*

**Department of Defense**
U.S. Army Corps of Engineers
20 Massachusetts Ave. NW
Washington, DC 20314
202-272-0010
*Oversees all construction of dams, reservoirs, bridges, harbors, and related projects; has principal responsibility for portion of Clean Water Act related to protecting wetlands.*

**Department of Energy**
Federal Energy Regulatory Commission
825 N. Capitol St.
Washington, DC 20426
202-357-8118
*Responsible for regulating electric utilities, including licensing of dams and hydroelectric projects.*

Office of Conservation and Renewable Energy
1000 Independence Ave. SW
Washington, DC 20585
202-586-9220
*Oversees all energy conservation programs, including research related to solar, wind, geothermal, and other renewable energy resources.*

**Department of the Interior**
Bureau of Land Management
Main Interior Building.
18th and C Streets NW
Washington, DC 20240
202-343-5717
*Manages public lands and mineral resources on public and some private lands in 10 western states and Alaska.*

Bureau of Reclamation
Main Interior Bldg.
18th and C Streets NW
Washington, DC 20240
202-343-4662
*Administers water and energy programs for public lands in 17 western states.*

National Park Service
Main Interior Bldg.
Washington, DC 20240
202-343-4747
*Sets policy for and manages all national parks.*

Office of Surface Mining
1951 Constitution Ave. NW
Washington, DC 20240
202-343-4953
*Oversees all strip-mining activities, including reclamation of mined land.*

U.S. Fish and Wildlife Service
Main Interior Bldg.
Washington, DC 20240
202-343-5634
*Responsible for conservation of fish and wildlife natural resources, including managing marine environmental quality.*

**Department of Transportation**
Coast Guard
2100 2nd St. SW
Washington, DC 20593
202-267-2229
*Oversees cleanup operations after oil spills, including inspections of vessels that carry oil and other hazardous substances.*

Federal Highway Administration
400 7th St. SW
Washington, DC 20590
202-366-0660
*Oversees all highway construction, including environmental impact of highway construction; regulates movement of hazardous wastes over U.S. highways.*

**Environmental Protection Agency**
401 M St. SW
Washington, DC 20460
202-382-2080
*EPA hotlines*: Drinking Water: 800-426-4791; Pesticides: 800-585-7378; SARA (Title III) Right-to-Know: 800-535-0202; Superfund: 202-382-3000, 800-424-9346.
*Responsible for enforcing all environmental policies, research, and regulations related to air pollution, water pollution, solid-waste disposal, hazardous-waste disposal, pesticide contamination, noise pollution, and radiation.*

## State Environmental Offices

Asterisk denotes each state's recycling coordinating agency

**Alabama**
Department of Conservation and Natural Resources
64 N. Union St.
Montgomery, AL 36130
205-261-3486

* Department of Environmental Management
Solid Waste Division
1751 Congressman W.L. Dickinson Dr.
Montgomery, AL 36130
205-271-7700

**Alaska**
* Department of Environmental Conservation
P.O. Box O
Juneau, AK 99811
907-465-2600

Department of Natural Resources
400 Willoughby
Juneau, AK 99801
907-465-2400

**Arizona**
Department of Environmental Quality
2005 N. Central Ave.
Phoenix, AZ 85004
602-257-2300

* Energy Office
1700 W. Washington St.
Phoenix, AZ 05007
602-542-3633

Land Department
1616 W. Adams St.
Phoenix, AZ 85007
602-255-4621

**Arkansas**
* Department of Pollution Control and Ecology
8001 National Dr.
P.O. Box 9583
Little Rock, AR 72219
501-562-7444

**California**
* Department of Conservation
Division of Recycling
1025 P St., Room 401
Sacramento, CA 95814
916-323-3743

The Environmental Affairs Agency
P.O. Box 2815
Sacramento, CA 95812
916-322-5840

The Resources Agency
1416 9th St., Room 1311
Sacramento, CA 95814
916-445-5656

**Colorado**
Department of Natural Resources
1313 Sherman, Room 718
Denver, CO 80203
303-866-3311

* Department of Health
Waste Management Division
4210 E. 11th Ave.
Denver, CO 80220
303-331-4830

Office of Energy Conservation
112 E. 14th Ave.
Denver, CO 80203
303-866-2507

**Connecticut**
Council on Environmental Quality
165 Capitol Ave., Room 239
Hartford, CT 06106
203-566-3510

* Department of Environmental Protection
Solid Waste Division
122 Washington St.
Hartford, CT 06106
203-566-5847

**Delaware**
* Department of Natural Resources and Environmental Control
89 Kings Hwy.
P.O. Box 1401
Dover, DE 19903
302-736-3869

**District of Columbia**
Conservation Services Division
613 G St. NW
Washington, DC 20004
202-727-4700

* Department of Public Works
Office of Policy and Planning
2000 14th St. NW, 7th Floor
Washington, DC 20009
202-939-8115

**Florida**
* Department of Environmental Regulation
2600 Blair Stone Rd.
Tallahassee, FL 32399
904-488-4805

Department of Natural Resources
Marjory Stoneman Douglas Bldg.
Tallahassee, FL 32303
904-488-1554

**Georgia**
* Department of Natural Resources
Floyd Towers East
205 Butler St.
Atlanta, GA 30334
404-656-3530

Institute of Natural Resources
University of Georgia
Ecology Bldg., Room 13
Athens, GA 30602
404-542-1555

**Hawaii**
* Department of Health
EPHS Division
P.O. Box 3378
Honolulu, HI 96801
808-548-6410

Department of Land and Natural Resources
Box 621
Honolulu, HI 96809
808-548-6550

Environmental Center
Water Resource Research Center
University of Hawaii
2550 Campus Rd.
Honolulu, HI 96822
808-948-7361

**Idaho**
* Department of Health and Welfare
450 W. State St., 3rd Floor
Boise, ID 83720
208-334-5879

Department of Lands
State Capitol Bldg.
Boise, ID 83720
208-334-3280

Department of Water Resources
1301 N. Orchard
Boise, ID 83720
208-327-7900

**Illinois**
Department of Conservation
Lincoln Tower Plaza
524 S. 2nd St.
Springfield, IL 62706
217-782-6302

* Department of Energy and Natural Resources
325 W. Adams St., Room 300
Springfield, IL 62704
217-785-2800

Illinois Environmental Protection Agency
2200 Churchill Rd.
Springfield, IL 62706
217-782-3397

**Indiana**
Department of Natural Resources
606 State Office Bldg.
Indianapolis, IN 46204
317-232-4020

* Department of Environmental Management
105 S. Meridian St.
P.O. Box 6015
Indianapolis, IN 46206
317-232-8603

**Iowa**
Department of Agriculture, Land Stewardship, and Division of Soil Conservation
Wallace State Office Bldg.
Des Moines, IA 50319
515-281-5851

* Department of Natural Resources
E. 9th and Grand Ave.
Wallace Bldg.
Des Moines, IA 50319
515-281-5145

**Kansas**
State Conservation Commission
109 SW 9th St., Room 300
Topeka, KS 66612
913-296-3600

* State Department of Health and Environment
Landon State Office Bldg.
900 SW Jackson St.
Topeka, KS 66612
913-296-1500

**Kentucky**
* Division of Waste Management
18 Reilly Rd.
Frankfort, KY 40601
502-564-6716

Environmental Quality Commission
18 Reilly Rd., Ash Annex
Frankfort, KY 40601
502-564-2150

Natural Resources and Environmental Protection Cabinet
Capital Plaza Tower, 5th Floor
Frankfort, KY 40601
502-564-3350

**Louisiana**
* Department of Environmental Quality
Solid Waste Division
438 Main St.
Baton Rouge, LA 70804
504-342-1216

State Office of Conservation
P.O. Box 94275
Capitol Station
Baton Rouge, LA 70804
504-342-5540

**Maine**
Department of Conservation
State House Station, #22
Augusta, ME 04333
207-289-2211

* Office of Waste Reduction and Recycling
286 Water St.
State house Station 154
Augusta, ME 04333
207-289-5300

State Soil and Water Conservation Commission
Deering Bldg.
AHMI Complex, Station #28
Augusta, ME 04333
207-289-2666

**Maryland**
Department of Natural Resources
580 Taylor Ave.
Annapolis, MD 21401
301-974-3987

* Department of the Environment
2500 Broening Hwy.
Baltimore, MD 21224
301-631-3000

**Massachusetts**
* Department of Environmental Protection
Division of Solid Waste Disposal
1 Winter St., 4th Floor
Boston, MA 02108
617-292-5961

Executive Office of Environmental Affairs
Leverett Saltonstall Bldg.
100 Cambridge St.
Boston, MA 02202
617-727-9800

**Michigan**
* Department of Natural Resources
Box 30028
Lansing, MI 48909
517-373-1220, TDD 517-335-4623

Water Resources Commission
P.O. Box 30028
Lansing, MI 48909
517-373-1949

**Minnesota**
* Department of Natural Resources
500 Lafayette Rd.
St. Paul, MN 55155
612-296-6157

Pollution Control Agency
520 Lafayette Rd.
St. Paul, MN 55155
612-296-6300

**Mississippi**
Bureau of Land and Water Resources
Southport Mall
P.O. Box 10631
Jackson, MS 39209
601-961-5200

* Bureau of Pollution Control
Department of Natural Resources
P.O. Box 10385
Jackson, MS 39209
601-961-5171

**Missouri**
Department of Conservation
P.O. Box 180
Jefferson City, MO 65102
314-751-4115

* Department of Natural Resources
P.O. Box 176
Jefferson City, MO 65102
314-751-3332

**Montana**
Department of Natural Resources and Conservation
1520 E. 6th Ave.
Helena, MT 59620
406-444-6699

Environmental Quality Council
State Capitol
Helena, MT 59620
406-444-3742

* Department of Health and Environmental Sciences
Cogswell Bldg.
Capitol Station
Helena, MT 59620
406-444-2544

**Nebraska**

* Department of Environmental Control and Land Quality
State House Station
Box 98922
Lincoln, NE 68509
402-471-2186

Nebraska Natural Resources Commission
301 Centennial Mall S.
P.O. Box 94876
Lincoln, NE 68509
402-471-2081

**Nevada**

Department of Conservation and Natural Resources
Capitol Complex, Nye Bldg.
201 S. Fall St.
Carson City, NV 89710
702-885-4360

* Office of Community Services
Capitol Complex #116
Carson City, NV 89710
702-885-4908

**New Hampshire**

Council on Resources and Development
Office of State Planning
2 1/2 Beacon St.
Concord, NH 03301
603-271-2155

* Department of Environmental Services
Waste Management Division
6 Hazen Dr., #8518
Concord, NH 03301
603-271-3503

State Conservation Committee
Department of Agriculture
Caller Box 2042
Concord, NH 03302
603-271-3576

**New Jersey**

* Department of Environmental Protection
401 E. State St., CN 402
Trenton, NJ 08625
609-292-2885

**New Mexico**

Environmental Improvement Division
1190 Saint Francis Dr.
Santa Fe, NM 87503
505-827-2850

* Health and Environment Department
1190 St. Francis Dr.
Santa Fe, NM 87503
505-827-2780

**New York**

* Department of Environmental Conservation
50 Wolf Rd.
Albany, NY 12233
518-457-5400

Environmental Protection Bureau
State Dept. of Law
120 Broadway
New York, NY 10271
212-341-2446

**North Carolina**
* Department of Human Services
Solid Waste Management Branch
P.O. Box 2091
Raleigh, NC 27602
919-733-0692

Department of Natural Resources and Community Development
P.O. Box 27687
Raleigh, NC 27611
919-733-4984

**North Dakota**
* Department of Health
Division of Waste Management
Box 5520
Bismarck, ND 58502
701-224-2366

Institute for Ecological Studies
P.O. Box 8278, University Station
University of North Dakota
Grand Forks, ND 58202
701-777-2851

**Ohio**
* Department of Natural Resources
Fountain Square
Columbus, OH 43224
614-265-6886

Environmental Protection Agency
P.O. Box 1049
1800 Watermark Dr.
Columbus, OH 43266
614-644-3020

**Oklahoma**
Conservation Commission
2800 N. Lincoln Blvd., Suite 160
Oklahoma City, OK 73105
405-521-2384

* Department of Health
P.O. Box 53551
Oklahoma City, OK 73152
405-271-7159

**Oregon**
* Department of Environmental Quality
811 SW 6th Ave.
Portland, OR 97204
503-229-5696

Water Resources Department
3850 Portland Rd. NE
Salem, OR 97310
503-378-3739

**Pennsylvania**
* Department of Environmental Resources
Fulton Bldg., 9th Floor
Box 2063
Harrisburg, PA 17120
717-787-1323

State Conservation Commission
Department of Environmental Resources
Executive House
P.O. Box 2357
2nd and Chestnut Streets
Harrisburg, PA 17120
717-787-5267

**Rhode Island**
* Department of Environmental Management
9 Hayes St.
Providence, RI 02908
401-277-2774

State Water Resources Board
265 Melrose St.
Providence, RI 02907
401-277-2217

**South Carolina**
* Department of Health and Environmental Control
J. Marion Sims Bldg.
2600 Bull St.
Columbia, SC 29201
803-734-5000

Division of Energy, Agriculture, and Natural Resources
1205 Pendleton St., Suite 333
Columbia, SC 29201
803-734-1740

**South Dakota**
Board of Minerals and Environment
Joe Foss Bldg.
Pierre, SD 57501
605-773-3151

Department of Water and Natural Resources
Joe Foss Bldg.
Pierre, SD 57501
605-773-3151

* Governor's Office of Energy Policy
217 1/2 W. Missouri
Pierre, SD 57501
605-773-3603

**Tennessee**
Department of Conservation
701 Broadway
Ellington Agricultural Center
Nashville, TN 37204
615-360-0103

* Department of Health and Environment
701 Broadway, 4th Floor
Nashville, TN 37219
615-741-3424

Energy, Environment, and Resources Center
University of Tennessee
327 S. Stadium Hall
Knoxville, TN 37996
615-974-4251

**Texas**
* Division of Solid Waste Management
1100 W. 49th St.
Austin, TX 78756
512-458-7271

State Soil and Water Conservation Board
P.O. Box 658
Temple, TX 76503
817-773-2250

Texas Conservation Foundation
P.O. Box 12845, Capitol Station
Austin, TX 78711
512-463-2196

**Utah**
* Bureau of Solid and Hazardous Waste
288 N. 1460 W.
Salt Lake City, UT 84116
801-538-6170

State Department of Natural Resources
1636 W. North Temple
Salt Lake City, UT 84116
801-538-3156

**Vermont**
Agency of Natural Resources
103 Main St.
Waterbury, VT 05677
802-244-7347

**Virginia**
Council on the Environment
903 9th St. Office Bldg.
Richmond, VA 23219
804-786-4500

Department of Conservation and Historic Resources
Division of Parks and Recreation
203 Governor St., Suite 306
Richmond, VA 23219
804-786-2121

* Division of Litter Control and Recycling
101 N. 14th St.
James Monroe Bldg., 11th Floor
Richmond, VA 23219
804-225-2667

**Washington**
* Department of Ecology
Olympia, WA 98504
206-459-6000

Department of Natural Resources
Public Lands Bldg.
Olympia, WA 98504
206-753-5327

**West Virginia**
* Conservation, Education, and Litter Control
1900 Kanawha Blvd. E.
Bldg. 3, Room 732
Charleston, WV 23505
304-348-3370

Department of Natural Resources
1800 Washington St. E.
Charleston, WV 25305
304-348-2754

**Wisconsin**
* Department of Natural Reources
Box 7921
Madison, WI 53707
608-266-2621

Land and Water Resources Bureau
Department of Agriculture, Trade, and Consumer Protection
801 W. Badger Rd.
Madison, WI 53713
608-267-9788

Wisconsin Conservation Corps
20 W. Mifflin, Suite 406
Madison, WI 53703
608-266-7730

**Wyoming**
* Environmental Quality Department
122 W. 25th St., 4th Floor
Herschler Bldg.
Cheyenne, WY 82002
307-777-7937

State Conservation Commission
2219 Carey Ave.
Cheyenne, WY 82002
307-777-7323